ARWED CRÜGER

Bargaining Theory and Fairness

Volkswirtschaftliche Schriften

Begründet von Prof. Dr. Dr. h. c. J. Broermann †

Heft 527

Bargaining Theory and Fairness

A Theoretical and Experimental Approach
Considering Freedom of Choice and the Crowding-out
of Intrinsic Motivation

By

Arwed Crüger

Duncker & Humblot · Berlin

Der Fachbereich Wirtschaftswissenschaften
der Martin-Luther-Universität Halle-Wittenberg hat diese Arbeit
im Jahre 2000 als Dissertation angenommen.

Bibliografische Information Der Deutschen Bibliothek

Die Deutsche Bibliothek verzeichnet diese Publikation in
der Deutschen Nationalbibliografie; detaillierte bibliografische
Daten sind im Internet über <http://dnb.ddb.de> abrufbar.

Alle Rechte vorbehalten
© 2002 Duncker & Humblot GmbH, Berlin
Fotoprint: Color-Druck Dorfi GmbH, Berlin
Printed in Germany

ISSN 0505-9372
ISBN 3-428-10741-1

Gedruckt auf alterungsbeständigem (säurefreiem) Papier
entsprechend ISO 9706 ♾

List of Contents

A. Introduction 15
    I. Motivation and Research Objectives 15
    II. Overview and Contents 17

B. Research on Bargaining Games 19
    I. Bargaining Games and Related Games 19
        1. The Ultimatum Game 20
        2. The Dictator Game 20
        3. The Impunity Game 21
        4. The Cardinal Impunity Game 21
        5. Cardinal Ultimatum Games 22
        6. The Best Shot Game and the Best Shot Mini Game 23
        7. An Auction Market Game 23
        8. Prisoner's Dilemma 24
    II. Comparisons Between Related Types of Games 24
        1. Experimental Results on Ultimatum and Dictator Bargaining 26
        2. Basic and Advanced Designs for Ultimatum Experiments 27
        3. Experiments with Dictator Games and Other Games 31
    III. A Summary of Research Results 36

C. Fairness and Intrinsic Motivation 41
    I. The Concept of Intrinsic Motivation 41
    II. Experimental Approaches Towards Intrinsic Motivation 42
    III. Aspects of Fairness 42
    IV. Referring to a Fairness Norm 44
    V. Relevant Factors for a Social Norm of Fairness 48

    1. The Level of Competitiveness.. 48

    2. The Level of Social Distance 49

    3. Annoyance as a Key Factor... 50

    4. Determinants for a Level of Annoyance.. 52

  VI. Another Implementation of Fairness... 53

  VII. Putting the Factors Together 55

D. Freedom of Choice.. 58

  I. The Basic Concept... 58

    1. Instrumental and Intrinsic Importance 58

    2. Negative and Positive Freedom 59

    3. Alternative Spaces, Functionings, and Capabilities 60

    4. The Famine Example 60

  II. Axiomatic Modeling of Freedom of Choice... 63

  III. Modeling Freedom of Choice with a Simple Game 63

  IV. A Summary on Freedom of Choice 64

E. The Two Games and Their Experimental Realization............................... 66

  I. Freedom to Punish... 68

    1. The Structure of the Game 68

    2. The Game Theoretic Solution of the FTP Game 70

  II. Right and Choice to Punish... 71

    1. The Structure of the Game 72

    2. The Game Theoretic Solution of the RAP Game 73

  III. Differences and Similarities Between the Two Games 74

  IV. The Experimental Realization.. 74

  V. The Experimental Procedure.. 76

F. Experimental Design for the FTP Game... 77

  I. Design Approach for the Experiment 78

    1. Treatment Variables.. 79

    2. Designs with a Low Proportional Bonus: A and B 79

List of Contents

    3. The Design Without a Bonus: C 80

    4. The Design with a Low Constant Bonus: D 81

    5. The Design with a High Constant Bonus: E 82

    6. Designs with a Constant Price: F, G and H 82

  II. Alternative Designs 83

G. Experimental Results for the FTP Game 85

  I. An Overview of the Decisions in the FTP Game 85

    1. The Veto Power Decisions 85

    2. The Proposals 87

    3. The Acceptance Decisions 93

    4. Payoffs and Efficiency 95

  II. Design Background and Hypothesis Approach 96

  III. Statistical Analysis for the FTP Game 97

    1. The Veto Power Decisions 97

      a) General Tendencies for the Veto Power Decisions 97

      b) Analysis of the Veto Power Decisions 98

      c) Graphical Illustration of the Veto Power Decisions 103

    2. The Proposals 105

      a) General Tendencies for the Demand Decisions 105

      b) Analysis of the Demand Decisions 106

      c) Graphical Illustration of the Demand Decisions 110

    3. The Acceptance Decisions 111

  IV. General Results of the FTP Game 113

    1. Interpretation of the Behavior Towards Freedom of Choice 113

    2. Overall Outcomes of the FTP Game 114

H. Experimental Design for the RAP Game 116

  I. Design Approach for the Experiment 117

    1. Treatment Variables 118

    2. Design I with a Small Bonus, a Fair and a Greedy Distribution 119

        3. Design II with a Small Bonus, a Greedy and a Very Greedy Distribution 121

        4. Design III with a High Bonus, a Fair and a Greedy Distribution 122

        5. Design IV with a High Bonus, a Greedy and a Very Greedy Distribution.... 123

        6. Playing a Subgame 124

    II. Alternative Designs 125

I. Experimental Results for the RAP Game 127

    I. An Overview of the Decisions in the RAP Game 127

        1. The Veto Power Decisions 127

        2. The Proposals 129

        3. The Acceptance Decisions 133

        4. The Subgames 135

        5. Behavior Types for Proposers and Receivers 136

        6. A Strategy Tournament 139

        7. Payoffs and Efficiency 144

    II. Design Background and Hypothesis Approach 146

    III. Statistical Analysis for the RAP Game 147

        1. Differences Between the FTP Game and the RAP Game 148

        2. The Veto Power Decisions 149

        3. The Proposals 151

        4. The Acceptance Decisions 158

        5. The Subgames 159

    IV. General Results of the RAP Game 161

        1. Interpretation of the Behavior Towards a Crowding-Out 161

        2. Overall Outcomes of the RAP Game 162

J. Summary 164

Bibliography 166

Subject Index 174

List of Figures

Figure 1: Suggested Average First-Round Ultimatum Demands 39
Figure 2: Determinants of Individual Behavior 57
Figure 3: Income and Life Expectancy in Five Developing Countries 61
Figure 4: Game-Tree with Parameters for the FTP Game 69
Figure 5: Payoffs for Demands of C and C - ε 70
Figure 6: Game-Tree with Parameters for the RAP Game 73
Figure 7: Session Overview for Both Games FTP and RAP 75
Figure 8: Design Overview for the FTP Game 78
Figure 9: Design Structure for the FTP Game 79
Figure 10: Game-Tree for Designs A and B with a Bonus δ of 10 % 80
Figure 11: Game-Tree for Design C with a Bonus of Zero 81
Figure 12: Game-Tree for Design D with a Bonus δ of DM 0,50 82
Figure 13: Realized and Expected VP Decisions for all FTP Designs 86
Figure 14: Veto Power Decisions for all FTP Designs 86
Figure 15: Expected Veto Power Decisions for all FTP Designs 87
Figure 16: Demands for all FTP Designs 88
Figure 17: Distribution of Demands for all FTP Designs 89
Figure 18: Expected Demands for all FTP Designs 90
Figure 19: Distribution of Expected Demands for all FTP Designs 91
Figure 20: Average Demanded and Expected Shares 92
Figure 21: Average Expected Demands for VP and NV Choices 92
Figure 22: Table of Acceptance Decisions for all FTP Designs 93
Figure 23: Acceptance Decisions for all FTP Designs 94
Figure 24: Expected Acceptance Decisions for all FTP Designs 95
Figure 25: Average Payoff Overview (in DM) 96
Figure 26: Veto Power Choices for all FTP Designs 98
Figure 27: Test Results for Hypothesis H^0_{FTP1} 100

List of Figures

Figure 28: Percentages of NV Choices for all Bonus Types 105
Figure 29: Average Demands for Designs B to H 106
Figure 30: Average Demands for all Bonus Types 110
Figure 31: Distribution of Rejected Offers 112
Figure 32: Design Overview for the RAP Game 117
Figure 33: Design Structure for the RAP Game 119
Figure 34: Payoff Table for Design I 120
Figure 35: Game-Tree with Parameters of Design I with $\delta = 5\ \%$ 120
Figure 36: Payoff Table for Design II 121
Figure 37: Game-Tree with Parameters of Design II with $\delta = 5\ \%$ 121
Figure 38: Payoff Table for Design III 122
Figure 39: Game-Tree with Parameters of Design III with $\delta = 50\ \%$ 122
Figure 40: Payoff Table for Design IV 123
Figure 41: Game-Tree with Parameters of Design IV with $\delta = 50\ \%$ 123
Figure 42: Game-Tree for Design ID 124
Figure 43: Game-Tree for Design IU 125
Figure 44: Realized and Expected VP Decisions for all RAP Designs 128
Figure 45: Veto Power Decisions for all RAP Designs 128
Figure 46: Expected Veto Power Decisions for all RAP Designs 129
Figure 47: Proposal Decisions for all RAP Designs 129
Figure 48: Demands in the Veto Power Situation 130
Figure 49: Demands in the Situation Without Veto Power 131
Figure 50: Expected Proposal Decisions for all RAP Designs 131
Figure 51: Expected Demands in the Veto Power Situation 132
Figure 52: Expected Demands in the Situation Without Veto Power 132
Figure 53: The Acceptance Decisions 133
Figure 54: Acceptance Decisions for High Demands 134
Figure 55: Expected Acceptance Decisions for High Demands 134
Figure 56: The Expected Acceptance Decisions 135
Figure 57: Proposal Decisions for all RAP Subgame Designs 135
Figure 58: The Acceptance Decisions 136
Figure 59: The Expected Acceptance Decisions 136

List of Figures

Figure 60: Behavior Types for the Proposer 138
Figure 61: Behavior Types for the Receiver 139
Figure 62: Strategies of the Proposer 140
Figure 63: Strategies of the Receiver 141
Figure 64: Favorite Strategies for Proposers 143
Figure 65: Favorite Strategies for Receivers 143
Figure 66: Average Payoffs for Proposers 144
Figure 67: Average Payoffs for Receivers 145
Figure 68: Efficiency of the RAP Designs 146
Figure 69: High Demands in the Absence of Veto Power 155
Figure 70: High Demands in the Veto Power Situation 156
Figure 71: Percentages of Equal Splits 157
Figure 72: Percentages of Rejections 158
Figure 73: High Demands in Designs I, II, IU, and IIU 160
Figure 74: High Demands in Designs I, II, ID, and IID 160

List of Abbreviations

e.g.	exempli gratia (for example)
etc.	et cetera
ERC	Equity, reciprocity, and competition
FGH	Fair-greedy, high bonus
FGHD	Fair-greedy, high bonus, Dictator version
FGHU	Fair-greedy, high bonus, Ultimatum version
FGS	Fair-greedy, small bonus
FGSD	Fair-greedy, small bonus, Dictator version
FGSU	Fair-greedy, small bonus, Ultimatum version
FTP	Freedom to Punish
GNP	Gross National Product
i.e.	id est
MWU	Mann-Whitney-U
RAP	Right and Choice to Punish
VGH	Greedy-very greedy, high bonus
VGHD	Greedy-very greedy, high bonus, Dictator version
VGHU	Greedy-very greedy, high bonus, Ultimatum version
VGS	Greedy-very greedy, small bonus
VGSD	Greedy-very greedy, small bonus, Dictator version
VGSU	Greedy-very greedy, small bonus, Ultimatum version
vs.	versus
w/o	without

List of Variables and Design Parameters

C	Cake
χ^2	Chi-Square for the Chi-Square-Test
δ	Bonus in the FTP Game, low bonus in the RAP Game
δ^*	RAP Game
ε	Smallest possible unit, i.e. DM 0,01
FTP	Freedom to Punish
H^0_{FTPX}	Null hypothesis No. X for the FTP game
H^A_{FTPX}	Alternative hypothesis No. X for the FTP game
H^0_{RAPX}	Null hypothesis No. X for the RAP game
H^A_{RAPX}	Alternative hypothesis No. X for the RAP game
L	Leave
N, NV	No veto power
p	In general: probability, here: significance level
P	Proposer
PY...	Strategies of the Proposer, explained in chapter I.I.6.
R	Receiver
RAP	Right and Choice to Punish
RN..., RV...	Strategies of the Receiver, explained in chapter I.I.6.
S^*	Complete equilibrium strategy for both players P and R
S_P^*	Complete equilibrium strategy for player P
S_R^*	Complete equilibrium strategy for player R
σ	Standard deviation of single values
T	Take
V, VP	Veto power
X	Number of a hypothesis
y	Demand in the FTP Game

List of Variables and Design Parameters

y, y^*	Low Demand in the RAP Game
Y, Y^*	High Demand in the RAP Game
Δy	Difference between high and low demand
Z	Z-value for Mann-Whitney-U statistical test

A. Introduction

The field of experimental economics has grown steadily for about half a century, providing models and studies that inspired even more research. For the history of experimental economics, see Kagel and Roth (1995); for the fundamental methods, see Davis and Holt (1993). In this study, two new bargaining situations are modeled and the collected experimental data is analyzed and interpreted. Along with the experimental findings, new theoretical concepts are considered and applied. Among these are the general concept of fairness, the crowding-out of intrinsic motivation and freedom of choice, all of which are not yet included in standard economic theory and therefore might prove to be worthwhile possible enhancements. This study supports the relevance of all of these concepts, and suggests some implementations and consequences.

I. Motivation and Research Objectives

Many economic models have been developed, then radically criticized, and finally refined, most of them for uncountable many times. The area of experimental economics offers another possibility. Instead of constructing models using just pure theory, experimenters are able to build models guided by existing laboratory data (Bolton 1998). These models can be easily and exhaustively tested by using new or more sophisticated laboratory methods, providing an instant and qualitatively controlled feedback.

Much of the innovative theoretical work has been inspired by the huge collection of experimental data or game theoretic approaches that were built up over the last decades. Some examples are the idea of "relative money" as an indicator for fairness (Bolton 1991), the game-theoretic modeling of fairness (Rabin 1993) or new functional forms of preferences (Bolton and Ockenfels 1999, Fehr and Schmidt 1999). The norm of fairness plays an important role in this context. The underlying intrinsic motivation, be it for fairness or other socially desirable norms, has to be defined, isolated and (if possible) measured (see Frey 1997c).

For example, the game theoretic prediction for Ultimatum bargaining experiments proved to be inaccurate to explain behavior in the laboratory. Two major observations could be extracted. Responders turned down meager but positive offers, therewith giving up money. And proposers made fair offers instead of using their strategic advantage, also giving up money. An equivalent observation holds even for the Dictator game, when Dictators offered sustainable amounts of money to the Recipients. Since this behavior does not maximize the payoff of the respective individual, it is not in line with the game theoretic prediction. Therefore, the underlying model of human motivation, which is based on monetary incentives according to standard economic theory, had to be enhanced. The baseline of all enhancements was fairness. The ERC theory of Bolton and Ockenfels (1999) takes fairness into account by including the relative payoff standing into the individual's motivation function. This new approach is successful in organizing a lot of laboratory data, including Ultimatum and Dictator experiments, but fails to predict the punishments observed by Ahlert, Crüger, and Güth (2001) in their so-called Equal Punishment game. Therefore, further refinements have to be done, and the relevance of intrinsic motivation in relation to fairness has to be analyzed. The game to be developed in this study called "Right and Choice to Punish" serves this purpose. The experimental results confirm the relevance of fairness as well as intrinsic motivation, and that they can play an important role for economic outcomes. But they also prove that no straightforward concept for the observed behavior exists and that fairness can be steady as well as fragile, meaning that it can prevail for a short or a very long time. It is also shown that a crowding-out of this intrinsic motivation is possible and sometimes even very likely to happen. Furthermore, an influence of the institutional frame on behavior was observed, especially by means of a comparison between the "Right and Choice to Punish" game and the "Freedom to Punish" game.

A second and related game called "Freedom to Punish" is also newly developed and aimed at another theoretical concept, freedom of choice. Both games add a new dimension to the existing Ultimatum and Dictator Literature. A first decision step is included, a possibility for the responder to choose between a situation with veto power, just like in an Ultimatum game, and a situation without veto power, just like in a Dictator game. This might make the game more complex, but provides unique opportunities to observe characteristics of both concepts, intrinsic motivation as well as freedom of choice.

In contrast to standard choice theory, freedom of choice – very roughly – assigns positive values to all kinds of alternatives, be it wanted or unwanted choices. The pure existence of another alternative raises the freedom of choice of the respective individual and is therefore a welcome and enriching (new) possibility. During the past ten years an increasing number of authors have modeled the individual welfare that arises from having the freedom to choose

A. Introduction

from a given set of alternatives. Several sets of axioms have been proposed to characterize rankings of opportunity sets in terms of freedom of choice. The experimental investigation to be studied in this work may contribute to the existing research in that area. Therefore, several experiments were conducted with "Freedom to Punish", a game that is a combination of a Dictator game (a no-choice-situation for the receiver) and an Ultimatum bargaining game (the receiver can choose between the two options accept and reject). The objective of the analysis is to investigate whether receivers prefer to have some freedom of choice or to have no choice dependent on the size of the monetary payoffs. The experimental results strongly support the idea of freedom of choice: players were not willing to drop an alternative without incentives to do so, but even with a small bonus they gave up their freedom of choice and excluded this alternative. As might be expected, higher monetary incentives generated more exclusions. The structure of this study is illustrated in the following paragraph.

II. Overview and Contents

The theory of bargaining has always been one of the main areas of interest for experimental economists, and therefore the existing results are both numerous and very diversified. Chapter B. summarizes the theoretical and experimental work in the field of bargaining. Even though some very helpful surveys by Güth and Tietz (1990), Roth (1995) or Güth (1995) already exist, an updated compilation is necessary since a great number of studies, which are especially relevant for this work, have been produced during the last couple of years. Furthermore, past results are grouped and analyzed to clarify how this study fits into existing research and theory. Grounding on that, it is shown how those research results may interact with the phenomenon of fairness. Therefore, chapter B. can also be seen as a short survey in the area of two person bargaining experiments.

Chapter C. deals with some aspects of fairness and intrinsic motivation, as well as with some of the theoretical work based on the existing amount of experimental data or game theoretic approaches. The connection to the underlying intrinsic motivation in some of the models is discussed. The concept of a crowding-out of intrinsic motivation is explained and discussed. The other major concept relevant to the present study is called freedom of choice, which is described in more detail in chapter D. The baseline for an axiomatic approach is outlined and the applicability of an experimental approach is shown.

With these new concepts in mind, two new games, called "Freedom to Punish (FTP)" and "Right and Choice to Punish (RAP)", are developed, analyzed and compared in chapter E. Furthermore, the procedure that was used for the experimental implementation of the two games is described. Building up on this, the experimental design for the game Freedom to Punish is illustrated by chapter F. Chapter G. contains the experimental results. The hypotheses are developed, discussed and tested. The central hypothesis is aimed at the importance of freedom of choice.

An identical order is kept for the game Right and Choice to Punish with chapters H. and I., starting with the experimental design. Here, the main hypotheses are focusing on the difference between the FTP game and the RAP game, and on the crowding-out of intrinsic motivation.

In a final summary, important outcomes for both games are highlighted and compared, and theoretical implications are discussed in chapter J. A closing outlook identifies areas for improvements and further research.

B. Research on Bargaining Games

For the vast body of literature on bargaining games, a survey like Roth (1995) provides a good overview. The probably most popular bargaining game is the Ultimatum game. Hundreds of experiments based on the Ultimatum game have been conducted, and the results were published in a huge number of articles. This literature has been surveyed by Güth and Tietz (1990), Güth (1995), and Roth (1995). Several recent articles like Bolton and Ockenfels (1999) also provide summaries or comparisons of important research results.

Every bargaining process could possibly end with an ultimatum. Therefore, this option has to be examined. And if ultimatums did not occur in certain bargaining situations, this should also be explained (Güth and Tietz 1990). A further reason for the popularity and also the importance of the Ultimatum game is that, despite its simplicity, most of the decisive features of bargaining can be taken into account. Güth (1995) points out "it's good to start from scratch with a basic model and think about enhancements later". Another advantageous feature of the simple game structure is that it makes experimental implementation a lot easier. But even though the Ultimatum game appears to be quite basic, studying it experimentally proved to be rather challenging.

The discussion about the Ultimatum game has also become a quest for the importance that fairness plays in bargaining. Learning and strategic reasoning also have to be considered when the behavior and the underlying decision process is analyzed and interpreted (Bolton 1998). A lot of new theories have evolved, presenting complex thoughts based on adaptive learning (Abbink, Bolton, Sadrieh, and Tang 1998), reinforcement learning (Erev and Roth 1998), relative payoff standing (Bolton and Ockenfels 1999) or inequity aversion (Fehr and Schmidt 1999). New game theoretic models were also developed and tested (Bolton 1991, 1993). Other studies propose new approaches towards the theoretical modeling of fairness (Rabin 1993). All this work is aimed at explaining behavior in the Ultimatum game or at least inspired by it.

I. Bargaining Games and Related Games

In the next chapters, some of the most important bargaining games are illustrated. Furthermore, some other games are also described, namely a market game, a dilemma game and a public good game. The reason for this approach is the collection of experimental results, which follows right after these game descriptions. A lot of experimenters have conducted experiments using two or more different kinds of games and compared the outcomes of the respective games. Therefore, the understanding of quite a variety of games is needed to follow these experimental designs. The focus is aimed at two person, one shot games, but some multiperiod and multiplayer games are also included if the design seems to be relevant for the compilation of experimental results, following after the game descriptions.

1. The Ultimatum Game

In the Ultimatum game, a first mover (proposer) proposes a division of a fixed monetary sum, called cake or pie, to a second mover (responder). If the second mover accepts, the money is divided accordingly; if he rejects, both players receive nothing. The subgame perfect solution, as described by Selten (1975), would presume that the proposer receives virtually the entire cake. Despite this plain solution, experimental researchers, starting with the work of Güth, Schmittberger and Schwarze (1982), report about considerable offers made by proposers, and on receivers rejecting non-zero offers. This motivated other experimenters to challenge these results, leading to the popular explanation that a fairness motive exists, therewith assuming that behavior is at least partly driven by a perfectly fair 50-50 division of the cake.

2. The Dictator Game

The Dictator game is even simpler than the Ultimatum game. One player, the dictator, decides how to distribute a fixed amount of money between himself and one other, the responder. The responder has no choice than to accept the proposal, and is therefore rather a recipient. Therefore, the Dictator game is more of a one person decision task than a real bargaining game, but it also provides some basic elements of bargaining: the underlying decision process of the proposer might be influenced by aspects of fairness and also by certain other psychological phenomena, for example crowding-out, see Bolton and Katok (1998), or kindness, see Bolton, Katok, and Zwick (1998). Therefore, it can't be analyzed by means of decision theory alone. Due to its few specifications, the Dictator game is not restricted to being a pure division task. Bolton and Katok (1998) interpret it as a public good game, and present a test on pure versus impure altruism. Again, the game theoretic analysis of the Dictator game leaves the Recipient with nothing, predicting that a perfectly rational Dictator is supposed to claim the whole cake for himself. Similar to the Ultimatum game, laboratory researchers found substantial giving by Dictators instead of pure selfish behavior, see Forsythe, Horowitz, Savin, and Sefton (1994), Hoffman, McCabe, Shachat, and Smith (1994), and also Bolton, Katok, and Zwick (1998).

3. The Impunity Game

To some extent, the Dictator game belongs to the group of Impunity games, since the dictator can never be punished. A different definition for Impunity games allows the responder to reject his payoff, while the payoff of the proposer remains untouched. All other features of the game are the same as in the Dictator game. To keep further analysis simple, the term "Impunity game" will only be applied to Impunity games that include this restricted rejection option. Such games without a rejection option will be called Dictator games.

The Impunity game is closer to a Dictator game than to an Ultimatum game, even though it includes a restricted rejection option. Such a rejection option is not given in the original Dictator game, but it also differs from the rejection possibilities in Ultimatum games, because the payoff of the Impunity Dictator is not questioned. Therefore, the second round decision in the Impunity game is somewhat arbitrary, since a rejection never reduces payoff inequality or relative payoff standing, but rather enlarges it. Nevertheless, it offers a way to the responder to show his frustration.

4. The Cardinal Impunity Game

The cardinal Impunity game was introduced by Bolton and Zwick (1995), and is also used by Bolton, Katok and Zwick (1998). In a cardinal Impunity game, the dictator can only chose between an equal division of the pie and a fixed division that favors the dictator but leaves both players with a positive payoff. Just like the regular Impunity game described in 3 above, the Recipient is allowed to reject the proposal of the Impunity Dictator and leave the game with nothing. Again, this does not affect the payoff of the dictator, as long as the unequal split was chosen. A rejection of the equal split leads to a payoff of zero for both players. This is in contrast to the (not cardinal) Impunity game of Güth and Huck (1997), where rejections never affected the payoff of the dictator, no matter if an equal or an unequal split was proposed. The game theoretic solution urges the Impunity Dictator to choose the unequal split. The cardinal character of this game offers some interesting design possibilities and compares to the game introduced later, "Right and Choice to Punish". Instead of giving the proposer the possibility to choose between all possible allocations, only two allocations are allowed. The advantages of this feature and the similar cardinal Ultimatum game are described in the next chapter.

5. Cardinal Ultimatum Games

So-called cardinal Ultimatum games (sometimes also named Ultimatum Mini games, Mini Ultimatum games or Reduced Form Ultimatum games) are studied by Bolton and Zwick (1995), but the basic idea of a game with a fair and an unfair offer was introduced by Güth and Yaari (1992), and analyzed by means of an evolutionary approach. Other game-theoretical work was done by Gale, Binmore and Samuelson (1995), while the experimental implementations of cardinal Ultimatum games include Güth, Huck, and Müller (1998), Abbink, Bolton, Sadrieh, and Tang (1998) and Abbink, Sadrieh, and Zamir (1999). The only difference to a normal Ultimatum game is the reduction to only two possible allocations, as already described in 4 above. Reducing the choice set to two allocations is likely to make technical and especially statistical analysis much easier and more convenient, but also offers additional advantages. Design variation offers another dimension of research, since the value of the unfair proposal can be varied, for example, between 60, 70, 80 and 90% of the cake. Each design then produces a simple and powerful experimental result.

The cardinal design offers another advantage regarding the questioning of subjects. The beliefs of the players can be obtained much easier, since a responder would usually have to produce a complete strategy considering all possible offers – and there are exactly 1.001 possibilities when splitting 1.000 Cents (10 Euro or 10 US Dollars). In the cardinal case, only two questions are necessary. These simplified cardinal games might also support comparability, since it seems less complicated to compare the results of cardinal Ultimatum games with cardinal Dictator or cardinal Impunity games, as long as the aim of the research study can still be targeted with respect to the cardinal restriction. All of the recent studies mentioned in this paragraph have shown that cardinal Ultimatum games capture the relevant Ultimatum game characteristics.

6. The Best Shot Game and the Best Shot Mini Game

The Best Shot game of Harrison and Hirshleifer (1989) is a Public Good game. The first player chooses an investment amount. The second player observes this amount and then also offers a certain contribution to the same public good. Only the maximum investment amount of both players' investments is finally realized. The Public Good is then provided, both players benefit in payoffs and only one player pays for it. The other player is a "freerider". There are two Nash equilibria: The first mover chooses the overall Pareto-optimal investment amount and the second mover zero, or the first mover contributes nothing and the second mover plays the Pareto investment.

The Best Shot Mini game of Gale, Binmore and Samuelson (1995) reminds of a cardinal Ultimatum game. Only two investment amounts are possible, zero and the "overall Pareto-optimal" investment. Of course, the latter is the only real investment and therefore the only option that would provide the Public Good and possibly lead to a positive payoff. In this situation, the first mover can take into account fairness (contributing to the Public Good) and own monetary payoff (free-riding). He can offer an investment or nothing. His choice might also heavily depend on his expectations about the other players' response, who can either "accept" or "reject" greedy behavior by investing the Pareto amount or also nothing. With respect to this choice set, the similarity between the Best Shot Mini game and the cardinal Ultimatum game is rather obvious.

7. An Auction Market Game

Roth, Prasnikar, Okuno-Fujiwara and Zamir (1991) describe a simple one-period auction market game. Only one seller offers one single unit of a good to nine buyers. The exchange of the good generates a fixed surplus. Each of the nine buyers submits one offer. The seller can accept or reject the best offer. The surplus is allocated depending on the market price. In any subgame perfect equilibrium, the seller receives nearly the whole surplus. Fairness would suggest a 50:50 split of the surplus given one seller and one buyer, but the competition between the existing nine buyers favors the seller and seems to crowd out fairness.

8. Prisoner's Dilemma

The Prisoner's Dilemma is probably the best known of all game-theoretic models and will therefore only be described very shortly. The Prisoner's Dilemma is not a bargaining game but a dilemma game. Both players have the same two strategies, defect and cooperate. Defecting is the dominant strategy and therefore the game theoretic prediction. This means that the Pareto-optimal solution, Cooperation, is not realized, resulting in a loss of efficiency. Therefore, an economic incentive to establish Cooperation exists. For an example see Bohnet and Frey (1999a). While the Prisoner's Dilemma is somewhat of the prototype of a dilemma game, Public Good games like the Best-Shot game also belong to the group of dilemma games. Bolton (1998) points out that the Prisoner's Dilemma is closely related to the Dictator game. Other similarities, for example between Dictator and Ultimatum game, are even more obvious. Therefore, comparisons of experimental outcomes of closely related games are common and often provide fascinating results.

II. Comparisons Between Related Types of Games

Comparing the experimental outcomes of certain games has become rather popular in the recent literature. A few of the results could be expected, but others might shed some new light on existing behavioral theory or recommend further enhancements. Some relevant comparisons are listed here, while the experimental results are shown in the next chapters.

Forsythe, Horowitz, Savin, and Sefton (1994) compare Ultimatum and Dictator games, as well as Hoffman, McCabe, Shachat, and Smith (1994). Suleiman (1996) also compares Ultimatum and Dictator games, but adds another innovative dimension of games by varying the veto power of the responder by using a discount factor. If the proposal is rejected by the responder, both payoffs are discounted by a factor δ with $0 \leq \delta \leq 1$, whereas $\delta = 1$ results in a Dictator game and $\delta = 0$ in an Ultimatum game. Therewith, this study not only compares Ultimatum and Dictator situations, but also allows observing the whole variation of games in between these two extremes.

Güth and Huck (1997) ran a design with a similar intention. They concentrated on four types of bargaining games, each with a different veto power for the responder. The Dictator game and the Impunity game have a low veto power, while the Ultimatum game and the (unnamed) fourth game have a high veto power. The veto power of the fourth game was actually the highest, because responders could reject the payoff of the proposer but keep their own payoff. Whether Impunity or Dictator games have a higher veto power is hard to answer. Since the responder is not able to harm the payoff of the proposer in both games, his respective position is probably equally weak.

The Equal Punishment game of Ahlert, Crüger and Güth (2001) can also be seen as a game settled somewhere in between Ultimatum and Dictator games, at least regarding the veto power of the responder. He can punish, which is differing from a Dictator game, but not as hard as in the Ultimatum game, since he is not allowed to reject the whole allocation as such. Instead, he can chose a punishment amount as high as his total payoff. This amount is deducted from both players' payoff, leading to zero payoffs for both players only in case of an equal division and maximum punishment. In case of an unfair division for the responder, he can reduce the payoff of the proposer only by reducing his own payoff as well.

Bolton and Zwick (1995) compare cardinal Ultimatum and cardinal Impunity games. Bolton, Katok and Zwick (1998) conduct cardinal Dictator games and compare the results with the outcomes of the cardinal Impunity games of Bolton and Zwick (1995). Gale, Binmore and Samuelson (1995) compare Ultimatum, cardinal Ultimatum and Best Shot Mini games. Prasnikar and Roth (1992) as well as Duffy and Feltovich (1999) analyze Ultimatum and Best Shot games. Bargaining games like the Ultimatum game and Public Good games like the Best Shot game have one thing in common – both are defining a certain distribution or division scheme. Bargaining games divide a pie, while Public Good games divide a burden to finance a public good. Therefore, comparing these games has become rather common in recent studies. Bolton and Katok (1998) interpret the Dictator game as a Public Good game and report on a crowding-out effect.

Auction Market games and Ultimatum games also prove to have a relation, since both produce a take-it-or-leave-it situation. Roth, Prasnikar, Okuno-Fujiwara, and Zamir (1991) compare the experimental outcomes of an Auction Market game with an Ultimatum game. Bohnet and Frey (1999a) compare the results of a Prisoner's Dilemma experiment with a Dictator game experiment. Some of the outstanding results of these comparative studies are illustrated in the following chapters along with other research outcomes.

1. Experimental Results on Ultimatum and Dictator Bargaining

The games under consideration and possible comparisons between games were already described. The major outcomes are listed next. After that, a selection of special issues is compiled, because the existing approaches are so divergent and complex that an only a more detailed description of selected areas can provide the necessary understanding of important design aspects and differences. The research in the area of Dictator and Ultimatum bargaining games has already advanced to a high level, leading to widely accepted general results. For example, the outcomes proved to be thoroughly stable as long as the experiments are conducted with comparable instructions, see Bolton, Katok and Zwick (1998). These stable results start with Güth, Schmittberger, and Schwarze (1982), and since then, a lot of replications with similar outcomes have been conducted.

Forsythe, Horowitz, Savin, and Sefton (1994) report on stable Dictator results in different time periods. These results are confirmed by Hoffman, McCabe, Shachat, and Smith (1994). Bolton, Katok, and Zwick (1998) demonstrate stable Dictator Giving despite several different experimental frames. Furthermore, the anonymity dispute seems to be settled. While Hoffman, McCabe, Shachat, and Smith (1994) claim an anonymity effect, Bolton, Katok, and Zwick (1998) clearly reject this explanation and attribute the observed irregularities to differences in the game frame. They also conduct a test for an anonymity effect and reject the existence of such an influence. For Ultimatum games, the anonymity effect was explored by Bolton and Zwick (1995), who found out that the willingness to punish provides a far better explanation for the observed behavior than an anonymity effect.

According to Bolton and Zwick (1995), two of the most distinctive behavioral regularities of the Ultimatum game, the rejection of unfair offers by responders and the consistently high frequency of fair offers by proposers, remain unchanged in Mini Ultimatum games. The implications of using the strategy method are discussed by Roth (1995) as well as by Güth, Huck, and Müller (1998). While the strategy method provides the experimenter with a broader range of data, a spontaneous play exhibits the participating subjects to a situation that might build up stronger emotions. The impact of envy or anger on behalf of the responder would be fully exposed in the spontaneous play scenario. Therefore, both options have certain advantages and disadvantages. For other important results and alternative interpretations, see Roth (1995).

2. Basic and Advanced Designs for Ultimatum Experiments

This chapter concentrates on experiments with a pure Ultimatum game and shows relevant results of selected designs. The following chapter summarizes Dictator games, mixed games and other related studies. For more detailed surveys – of course without recent studies – see Güth and Tietz (1990) as well as Roth (1995), or Güth (1995) for a personal review. The experimental implementations of Ultimatum bargaining start with Güth, Schmittberger and Schwarze (1982). They conducted experiments with several designs, and with the basic version being a simple one-shot Ultimatum game. This framing had cake sizes from DM 4 to DM 10 and produced the first experimental results for Ultimatum bargaining, which should prove to be typical and stable outcomes for this kind of game. The average demand was 65% of the cake, 33% of the offers were equal splits and 9.5% of all offers were rejected.

The studies of Binmore, Shaked, and Sutton (1985, 1988) are somewhat different in the aim of their work, and also in the rules of the game they are examining. Nevertheless, their two-stage game produced similar results, especially for the first stage. They show some differences, but have to admit that this might be produced by different conditions. About 36.6% equal split offers occurred and the average demand was 56.5%. Only a subsample of the response data is reported, with a rejection rate of 13.6%. The similarities are obvious.

Kahnemann, Knetsch, and Thaler (1986a, 1986b) perform experiments with students of psychology and students of business administration, and report deviating results. While some of the differences might be due to experimental conditions (see Güth and Tietz, 1990), they can clearly show that psychology students propose equal splits far more often (roughly 80%) than students of business administration (63%). Therewith, the proportion of equal splits was significantly higher than in other Ultimatum experiments, but a different experimental setting might as well have caused this. Nevertheless, the mean share allocated to responders was again substantial (45%), meaning that it was far higher than the game theoretic prediction of close to zero, but not exactly fair, since it is also below 50%.

Roth, Prasnikar, Okuno-Fujiwara, and Zamir (1991) show that the results for Ultimatum bargaining games is at least partly influenced by culture. Other culture effects are reported by Burlando and Hey (1997) as well as Ockenfels and Weimann (1999). Roth et al. (1991) point out that the observed difference between bargaining behavior in four different subject pools is not necessarily based on a different level of aggressiveness, but rather on different expectations about what would be a reasonable or fair offer.

A popular enhancement to Ultimatum bargaining is offered by Rubinstein (1982). In the Rubinstein bargaining game, a rejection does not lead to zero payoffs but to another round of bargaining, where the initial cake is decreased by a certain discount factor and the bargaining positions are exchanged between the two participants. Even though the Rubinstein model is constructed upon more than one period of bargaining, its experimental results are worth discussing. This setting is especially interesting because the existing discount factor can be interpreted towards the monetary loss caused by the time consumption of long lasting bargaining situations. Typical results are surveyed by Güth and Tietz (1990). Very roughly spoken, the first round offers are not dramatically different from usual Ultimatum bargaining behavior, and average at around two thirds of the cake. As expected, rejections are far more frequent, since the bargaining process continues into a next round in that case, except of course for the last round. Another interesting finding is the fact that a bigger cake leads to higher relative demands.

Another approach is using an increasing instead of a decreasing cake. Güth, Ockenfels, and Wendel (1993) show that even though game theory would induce Ultimatum offers in the first period, most (68.6%) of the plays did not produce an Ultimatum in Period one, instead bargaining was postponed into the next period, which of course increased efficiency due to a higher available cake. Additionally, they report a boundary for acceptance at exactly two thirds, which is remarkable since a unique boundary can rarely be identified and only average values are usually extracted. In this sample, all demands below two thirds are accepted, while all demands above are rejected.

Güth and Tietz (1990) found out that a bigger cake leads to higher relative demands in multiperiod Ultimatum games. This has also been tested for simple one-period Ultimatum games. Güth and Tietz (1990) also showed that responders are willing to accept a lower share if it is taken from a larger cake. Hoffman, McCabe, and Smith (1996a) varied the monetary stakes using 10 or 100 US Dollars and found a decreasing rejection rate in a pooled data sample, but no differences in the distributions of offers. Forsythe, Horowitz, Savin and Sefton (1994) also reported no differences in offer distributions for stake variations between 0 and 5 and between 5 and 10 US Dollars, but a lower rejection rate for higher stakes, just like the two other studies. Therefore, it comes as a bit of a surprise that Slonim and Roth (1998) claim to be "the first to detect a lower frequency of rejection when stakes are higher (p. 569)". They conducted experiments in the Slovak Republic using stakes of 60, 300 and 1500 Slovak Crowns, equivalent to 1.90, 9.70 and 48.40 US Dollars. Nevertheless, they might have been the first ones to observe that offers decrease when some experience is gained in the ten rounds of play for the high stakes condition.

II. Comparisons Between Related Types of Games

Different methods can be used to raise the quality of the decisions made by the participants of economic experiments. To make sure that all subjects read the instructions thoroughly, a pre-play phase, sometimes called decision preparation, can be conducted. The simplest example is a questionnaire. Participants are forced to answer questions about the game situation, and therefore have to explore and consider certain important aspects of the experimental setting before the real play phase actually begins. A lot of experimenters use questionnaires to determine whether subjects understand the game and are ready and willing to play it or not. But questionnaires can also be used as a treatment variable. Tietz (1992) shows that a decision-planning phase helps to form realistic expectations and reduces the reluctance to trade in a simple market situation.

Another possibility is auctioning the different game positions (see Güth and Tietz 1986), which should also enforce an intensive studying of the strategic possibilities by the players before starting the game. Additionally, the auction winners have to pay a price to gain their game position, and might therefore be concentrating harder on optimizing their monetary payoff, probably pushing the experimental results closer towards the game-theoretic solution. In fact, only one equal split (3%) occurred, but the average demanded shares (56% to 72%) do not differ dramatically from Ultimatum games without Auctions. The number of rejections is also in the previously observed range (11%). This slight shift towards the game-theoretic solution is confirmed by the results of Güth, Ockenfels, and Tietz (1992).

A similar approach is using a quiz to allocate the player's positions. When the different roles for the Ultimatum game are not assigned to the participating students at random, but based on their success in a knowledge quiz, the behavior in the game also tends to move towards the game-theoretic solution, see Hoffman, McCabe, Shachat, and Smith (1994) or Hoffman, McCabe, and Smith (1996a). Similar results are reached using a simple mathematical contest; see Hoffman and Spitzer (1982, 1985). Again, less equal splits are observed when positions are earned, and the share kept by proposers increased but did not reach the game-theoretic solution.

Instead of concentrating on how positions are achieved, one can also vary the way the cake is achieved by the subjects. Instead of giving a gift to participants, the experimenter could make the subjects work for their cake. In a production game, the cake is produced by both players and then the division of this cake is negotiated using an Ultimatum game, see Königstein and Tietz (1994), or for the reverse version with first-round Ultimatum bargaining about the division in percent and the production of the cake in the second round see Crüger (1996) as well as Crüger and Königstein (1999). The average demanded share for the proposer in the experimental results of Crüger (1996) was only

52%, probably caused by the fear to annoy the responder and therewith force him to provide a low production effort. This fear seemed to foster fair behavior among proposers, also indicated by the high proportion of equal splits (55%). Another explanation might be the complexity of the game, leading to only vague expectations. Overall, 11% of the offers were rejected. Königstein and Tietz (1994) observed 14% rejections, and a mean demand of 61% with 22% of the offers being equal splits, measured using the division of the generated surplus. The difference between these studies can be attributed to the different game structure, namely Ultimatum bargaining with advance production and, in the other case, with subsequent production.

The cardinal Ultimatum games of Bolton and Zwick (1995) show behavior consistent with the results of normal Ultimatum games. The number of equal splits was sustainable with 44%, and 20% of the offers were rejected. An average demand can only be estimated based on the data provided, and might be between 60 and 65%. Güth, Huck, and Müller (1998) apply some small changes to the game of Bolton and Zwick (1995). They use nearly equal splits instead of equal (fair) splits and find that nearly equal splits are made less often than equal splits in comparable conditions. Overall, their results are in line with previous reports, namely an average demand of 68% with 49% equal (or nearly equal) splits, and 18% rejections. Abbink, Bolton, Sadrieh, and Tang (1998) as well as Abbink, Sadrieh, and Zamir (1999) apply further changes and show effects on fairness and learning, but the general results regarding demands and rejections remain valid.

Straub and Murnighan (1995) suggest that small offers in Ultimatum games are rejected because the pride of the responders has been wounded, and this pride is restored by a rejection. They show that without knowing how much money is divided, i.e. without information about the size of the cake, a significantly higher portion of small offers is accepted. Furthermore, they suspect that expectations might have a considerable impact on agreements. This is confirmed by the study of Harrison and McCabe (1996), who manipulated the expectations of the players and therewith influenced their behavior. They suspect that considerations of fairness do not influence bargaining behavior independently of subject expectations.

A difference between individual and group behavior is reported by Bornstein and Yaniv (1998), where one group played the Ultimatum game with another group, each group consisting of three persons. While their general results concerning average demand (56%), equal splits (50%) and rejections (10%, all results taken from their double blind design No. 2) are again in line with previous findings, groups tended to play closer to the equilibrium prediction, since they offered less and were also willing to accept less than individuals.

The next study by Güth and van Damme (1998) explores a three player Ultimatum game, which is also a subtle combination of an Ultimatum and a Dictator game. A proposer offers a three-way split of the pie, and the responder can accept or reject, while the third player (actually a dummy) has no choice but to accept whatever is left for him, just like the receiver in the Dictator game. Usually, the amounts left for that third player by the proposer were marginal, and none of the responder's rejections could be explained by a low share for the receiver, which might have signaled some responder's concern for the well-being of receivers. Altogether, no fairness towards the helpless receiver could be observed, while responders usually received about one third of the cake.

Selten and Ockenfels (1998) introduce a similar idea with the solidarity game, where also three players form a group and each of them rolls a die and wins a certain amount of money if one of the winning numbers is up. But, since there might be some people in the same group of three who win and some who don't, participants are asked to offer a certain amount of money to be paid to possible losers before they actually roll. Therefore, every member of a group has to say how much he wants to tribute in case of one loser in his group, but also in the case of two losers. The striking finding is that most people give the same total amount for the case of one and for the case of two losers. Therefore, even though subjects generally seemed to be willing to give money, they were not really interested in a fair or nearly fair division among all three members of the group. Their major concern is clearly their own share of the cake.

The last two articles already dealt with situations in which a receiver has no choice but to accept what is offered to him. The behavior in situations without a rejection option has been observed in many Dictator game experiments, and these as well as other comparative studies are included in the next chapter.

3. Experiments with Dictator Games and Other Games

Even though game theory urges the Dictator to offer nothing to the recipient, the studies by Forsythe, Horowitz, Savin, and Sefton (1994), Hoffman, McCabe, Shachat, and Smith (1994), and also Bolton, Katok, and Zwick (1998) report on substantial giving. Some Dictators indeed leave nothing, but others give away as much as 50% of the pie. The modal amount given to the Recipients is sometimes as high as 30 percent. This clearly indicates that Dictator game giving cannot be expressed by a simple percentage figure, but instead is widely dispersed. An often-discussed explanation for this phenomenon targets experimenter observation. This hypothesis presumes that some

Dictators believe that the observing experimenter's assessment of them might be disadvantageous if they show the selfish greedy behavior that is predicted by game theory. To avoid such a reputation, Dictators are motivated to leave some of the cake to the Recipients. A previous study by Hoffman, McCabe, Shachat, and Smith (1994) finds evidence in support of this anonymity hypothesis. But it appears doubtful whether participants consider the assessment of the experimenter rather than their own self-perception or the thoughts and emotions of the fellow student they are interacting with. Another study by Bolton, Katok, and Zwick (1998) does not find any evidence for the anonymity hypothesis. Bolton and Zwick (1995) also consider this hypothesis, but do not find enough evidence, therefore favoring a totally different explanation focusing on the idea that unfair proposers are punished by responders. Generally, other explanations, for example based on a fairness motive, have to be clearly rejected before a somewhat artificial anonymity hypothesis should be taken into account. The Dictator game alone was obviously not challenging enough, because most of the experimenters compared their Dictator results with at least one control group that played another game, for example the Ultimatum or Best Shot game. This inspired other comparative studies, which are also listed below.

Hoffmann and Spitzer (1982, 1985) present an entitlement approach and develop a more complicated game to explore the Coase theorem. Their games include face-to-face communication and only a cooperative path leads to efficiency, but their results nevertheless show the same characteristics as other pure Dictator games, namely a not negligible share of equal splits. See also Harrison and McKee (1985).

Kahnemann, Knetsch, and Thaler (1986a, 1986b) perform Ultimatum and Dictator experiments and follow the entitlement idea of Hoffmann and Spitzer (1982, 1985). They use a totally different experimental approach based on telephone interviews. They argue that the right to punish in an Ultimatum game can be interpreted as an enforcement device. They also show that fair behavior happens in the absence of enforcement, i.e. in Dictator games.

Forsythe, Horowitz, Savin, and Sefton (1994) were the first to conduct both pure Ultimatum and pure Dictator game experiments using a pay and a no pay condition. They find that average offers for both games are clearly higher than the near zero subgame perfect prediction. While average shares between 19% and 25% were offered in the three Dictator game sessions, the Ultimatum game sessions produced offers between 44% and 47%. Equal splits dominated the Ultimatum game behavior with 52% to 71%, and still play an important but clearly smaller role in Dictator games with 14% to 21%. Rejections in Ultimatum games averaged 6%. While equity considerations might again have influenced behavior, the explanation is not that simple, because offers were considerably higher in the Ultimatum games.

Hoffman, McCabe, Shachat and Smith (1994) confirm the results of Forsythe, Horowitz, Savin, and Sefton (1994), and offer further insights by using a slightly enhanced version of the contest entitlement by Hoffman and Spitzer (1985), hereby introducing the right to be proposer as a property right. They show that this contest entitlement produces results closer to the game-theoretic solution for both Ultimatum and Dictator, but still a remarkable amount of money is offered by proposers. They also show that a slightly altered instruction text, formulating a buyer/seller exchange problem instead of a division task, shifts offers towards the equilibrium as well. Eckel and Grossman (1996a) built on the work of Hoffman, McCabe, Shachat and Smith (1994) and replicate their results. Furthermore, they replace the receiver in an additional design. The anonymous individuals are replaced with an established charity that becomes the receiver. This raised Dictator Giving from 10% to 30%.

Instead of a contest or quiz, Selten and Ockenfels (1998) used a die to determine winners (proposers) and losers (receivers), and therewith played a hidden three-player Dictator game, which was already described in chapter B. above. They also prove that some willingness to give money exists, since equal splits happen (depending on a strict or not so strict definition of equal splits, they average between 8% and 23%), and giving is also substantial with an average of 31% in case of two losers (DM 1,56 of the DM 5,- amount divided with each of the two losers) and an average of 25% in case of one loser. Furthermore, they find that the sum of giving matters, since subjects give the same total amount to one or to two losers.

This so called fixed sacrifice effect is also reported by Bolton, Katok, and Zwick (1998) in cardinal Dictator games. They also conduct Dictator games with six possible allocations instead of two, and with ten receivers instead of one. Dictator Giving tended to be the same total amount for one and for ten receivers on the basis of the same 10 dollar pie. When faced with more than one recipient, Dictators also showed no willingness to treat these recipients equally, which again raises doubts about the importance of egalitarian preferences. As already mentioned in paragraph B.II.1. above, they find no evidence for an anonymity effect, but propose that differences between previous results were caused by certain specialties in the experimental framing. They conclude that Dictator giving (an average of 14% of the pie) is primarily caused by a concern for a fair distribution on behalf of Dictators. But they also point out that, when faced with the choice to leave more or less than they would freely chose, Dictators most likely give less. Altogether, giving does not seem to be caused by concerns about the welfare of others, but by certain social rules that constrain pure self-interest. Nevertheless, within these constraints behavior is most likely driven by pure egoism.

Güth and van Damme (1998) also used a three player setting explained in chapter B. above, and the proposer faced a responder and a receiver. While the responder was treated much like in usual Ultimatum games with one-third offers, the receiver only obtained marginal amounts. This means that the sum of giving does not differ much from two-person Ultimatum bargaining proposals.

Bolton and Zwick (1995) conduct cardinal Impunity experiments and compare the outcomes with their cardinal Ultimatum results. Due to the close similarity to the Dictator game already explained in paragraph B.I.2. above, it comes as a surprise that Impunity Dictators play in accord with the equilibrium prediction, since they nearly never (only in 2% of all cases) leave more money than required. This is also in contrast to the cardinal Ultimatum offers, where only about one third were equilibrium offers. Therefore, they strongly support the punishment hypothesis – proposers avoid equilibrium offers because they fear being punished by a rejection. It seems to be worth noting that the cardinal character of these two games might possibly have caused this discrepancy, since the strategic advantage of the Impunity Dictator can only be exploited by choosing the greedy split, while in comparison to that the position of the Ultimatum proposer is weakened by the rejection option of the responder, and choosing the greedy split instead of the fair split bears a high risk of being punished. This issue will again be addressed later by the two games to be developed, FTP and RAP.

Other Dictator games are used by Bolton and Katok (1998) to explore crowding-out effects. They show again that giving can be higher due to a different instruction text. They suspect that a distribution task generates significantly different outcomes than a competitive game. They also find evidence for an extensive but incomplete crowding-out effect. A test for gender differences in Dictator Giving is reported by Bolton and Katok (1995). They find no evidence, but Eckel and Grossman (1996b) observed a gender effect in a Punishment game. So did the already mentioned study of Selten and Ockenfels (1998), where females showed far more solidarity than men.

Suleiman (1996) used a discount factor, already explained in chapter B.II. above, to vary from Ultimatum to Dictator games. Again, substantial portions of the cake were given to responders (respectively receivers) in all five discount settings, but the size of the offers strongly depended on the discount factor. The closer to a Dictator situation, the lower the offers. Furthermore, the equal split was the modal allocation for all treatments, and the percentage of equal splits also decreased depending on the discount factor.

II. Comparisons Between Related Types of Games

Güth and Huck (1997) compared four games with different veto power, namely the Ultimatum game, the Dictator game, the Impunity game and a fourth version with maybe the highest veto power, since the responder can reject the payoff of the proposer alone and keep his own monetary payoff. At least the results confirm this conjecture, since proposals where highest for the fourth game type. The second highest proposals were observed in the Ultimatum game, with a significant difference to the much lower offers in Dictator games. The Impunity game produced slightly lower offers than the Dictator game.

The Equal Punishment game of Ahlert, Crüger, and Güth (2001) is settled somewhere in between Ultimatum and Dictator games. The responder cannot reject, but punish both players with equal amounts that are deducted from the final payoffs. Since payoffs are not allowed to be negative, he can only punish using the money allocated to him plus his show-up fee. Therewith, the proposer can reduce possible punishment down to the show-up fee by offering nothing. Viewed in this light, the Equal Punishment game is up to some extent more likely to produce extremely unfair results than the Dictator game, since the proposer has to fear a higher punishment for i.e. a 10% offer (which would be in the range of some Dictator Giving mentioned in this paragraph) than for a zero offer. Nevertheless, totally fair equal split offers were frequent in the first round (32%), but nearly vanished in the repetition (6%). Average first round offers were about one third of the cake (without show-up fee), but also decreased in the repetition (13%). The results show that punishments are frequent, even though the payoff inequality is therewith expanded instead of reduced, as was the case with rejections of unfair offers in the Ultimatum game. Inequality aversion would indeed have predicted zero punishments. The importance of relative payoff standings is again questioned.

Among all comparisons other than Ultimatum versus Dictator (or closely related like Impunity) is a study by Roth, Prasnikar, Okuno-Fujiwara, and Zamir (1991). They compare Ultimatum Results, which were collected in four different countries, with the results of an Auction Market game, which was run in the same four countries. While the behavior across all four countries in the Auction Market game can be somewhat explained by the subgame perfect equilibrium, this does again not hold for the Ultimatum game data, where substantial offers and frequent rejections could be observed. It should be mentioned that the rules of the market as well as the instruction language might have enforced equilibrium play, compared with other possible market forms and instruction sets.

Based on the data of the study just mentioned, Prasnikar and Roth (1992) compare Best Shot games with Ultimatum games, and also find that the equilibrium is reached in Best Shot games, but not in Ultimatum games. The same holds for their comparison of Market game data and Ultimatum game data. Duffy and Feltovich (1999) build on the work of Prasnikar and Roth (1992) and also conduct Ultimatum and Best-Shot games. They concentrate on information and learning, but their general results are in line with previous studies. They find that equilibrium play is reached in Best Shot games, but not in Ultimatum games, where offers again averaged around one third of the cake.

Several Dictator experiments are compared by Hoffman, McCabe, and Smith (1996b). They show that the social distance between proposer and receiver has some significant influence on the distribution of offers. The less social distance exists, the more money is offered to receivers. They argue that other-regarding behavior might be triggered by the similarity to real life situations, and by expanding the social distance, self-interested actions become more likely. For a discussion of the impacts see the comment by Bohnet and Frey (1999b) as well as the reply by Hoffman, McCabe, and Smith (1999).

Bohnet and Frey (1999a) compare Prisoner's Dilemma and Dictator games. They find that a decrease of the social distance between subjects by means of a silent identification with each other significantly raises solidarity, expressed by equal splits in Dictator games and cooperative actions in the Prisoner's Dilemma. A similar approach is realized by Brosig, Ockenfels, and Weimann (1999), who show that cooperation in a public goods game increases with the intensity of pre-play communication, represented by several different communication mediums, while the content of communication, respectively the transferred information, is relatively stable along all treatments. Roth (1995) gives two possible explanations. The first is based on the assumption that intensive personal contact, i.e. face-to-face bargaining, might produce certain motivations, i.e. group identity, which arise from uncontrolled social aspects. The second thought targets the unlimited channels of communication that exist in face-to-face bargaining, and the resulting possibilities to avoid misunderstandings or efficiency losses. Altogether, the level of social distance and the existing communication possibilities should also be considered when experimental outcomes are analyzed or compared.

III. A Summary of Research Results

In the previous chapters, the huge experimental literature in the field of bargaining was browsed, and major results were highlighted. While some of these results will be directly compared with the results of this study, others rather served the purpose of demonstrating the stability of experimental outcomes under different framing conditions. The unexpected results of Güth,

Schmittberger and Schwarze (1982) challenged other authors to investigate and explain the underlying phenomenon, but as far as the main results are concerned, all of them only confirmed what was previously observed. Nevertheless, a lot of new approaches and enhancements were developed and tested, providing insights and conclusions about the interplay of certain aspects of the decision process.

The main findings were the existence of considerable and even perfectly fair equal split offers, and of frequent rejections of non-negligible amounts. It might be possible to present some overall results on a statistical basis, but such analysis should be concentrated on the new results in the second part of this work. Statistics would not be very productive here either, since a lot of designs are not easily comparable and too many differences would have to be handled. Furthermore, extensive statistics already exist, see Güth and Tietz (1990) or Roth (1995). Therefore, the following presentation of experimental results for offers and rejections is just a simplification without the necessary significance, but the underlying similarities between all kinds of designs are so obvious that it seems possible to refrain from such proof.

Maybe the most striking result of all of these studies is an average demand of about two thirds of the cake by proposers in Ultimatum games. Of course, this generalization appears to be dangerous because the range of these averages is between 55% and 72%. A very careful statement would only describe a high possibility for an average demand to be significantly more than 50% and less than 80%. Fehr and Schmidt (1999) aggregate ten studies and show that only 4% of all demands are above 80% of the cake and 71% of all demands are in a range between 50 and 60%, leaving most of the rest (nearly 25%, since only a few demands are less than 50%) in a range between 60 and 80%. But this leads nowhere. It is pretty common between experimenters to expect a two third average demand, and Güth, Ockenfels and Wendel (1993) even observe this demand to be the unique boundary for acceptance. This goes beyond average demands, since all offers below two thirds were accepted, and all above rejected. The interplay between proposers and responders is at least remarkable. Responders seem to expect at least one third of the cake or even anticipate the average proposer behavior of two third demands, and respond accordingly by rejecting offers that appear to be too far below their expected one-third share. These rejections seem to be anticipated by most of the proposers, who demand only two thirds (or less). The question is how subjects basically find such reference points like "two thirds of the cake". A trade-off between fairness and the strategic advantage of the proposer seems to exist, but how does it come about, and how is it being solved?

One of the problems of a two third simplification is of course the relatively rare occurrence of concrete two third demands. Usually, the modal offer is the equal split with 22% to 72% of all offers. This raises again the question of fairness and strategic reasoning. While some of the proposers take advantage of their position and demand more, others behave according to a norm of fairness and chose the equal split. Again, a fear of rejection is a possible explanation. And rejections indeed happen frequently, most studies report rejection rates between 10 and 20% (see also Roth 1995, p. 265). This is especially remarkable because the rejected amounts were often as high as 45% of the cake (or even more), which means that at least some of the subjects were willing to give up substantial amounts of money to punish greedy (and also slightly greedy) proposers.

Another interesting problem is the frequently observed effect of higher stakes on average demands or rejection frequency, as already mentioned in chapter B.II.2. above. The following Figure 1 is intended to serve as a suggestion for discussion, not as a new behavioral theory. It shows two curves, the average demand curve and the always accepted demand curve, both for first round Ultimatum bargaining behavior. The latter consists of the maximum demand that is accepted by all responders. To include all responders and therewith totally exclude the risk of a rejection, a demand should simply not be higher than 50%, and in certain situations even less, for example in the presence of very low monetary incentives. This can be explained by a perception problem. Responders do not accept amounts that are so low that there exists virtually no difference to a zero payoff. Thus, a rejection might not be caused by fairness considerations, but just because the offered amount is not of interest (Roth 1995), even though it might include some or even perfect fairness. Following this, an amount of DM 1 appears to be high enough to attract responders, also taking considerations of prominence (Tietz 1984) into account. Hence, a stake of DM 2 already produces an acceptable share for the responder when a 50% split is offered. For lower stakes than DM 2, the acceptance behavior is unpredictable without formulating other assumptions. To exclude any risk of a rejection, proposers have to demand below the fair 50% share, but the higher the stake and the higher the absolute payoff of the responder, the less likely is a rejection of 50%, which is the reason why the always accepted demands curve slowly approaches to 50% the higher the stakes are.

The development of the average demand curve follows similar considerations. Of course, there is a difference between average accepted and average rejected offers, which is also most likely to change with higher stakes. But since the share of rejections with 10 to 20% is not of dramatic influence, and proposers do most likely have some expectations about rejection behavior which are already included in their demands, the average demand curve still provides some useful insights, even though it includes accepted as well as

III. A Summary of Research Results

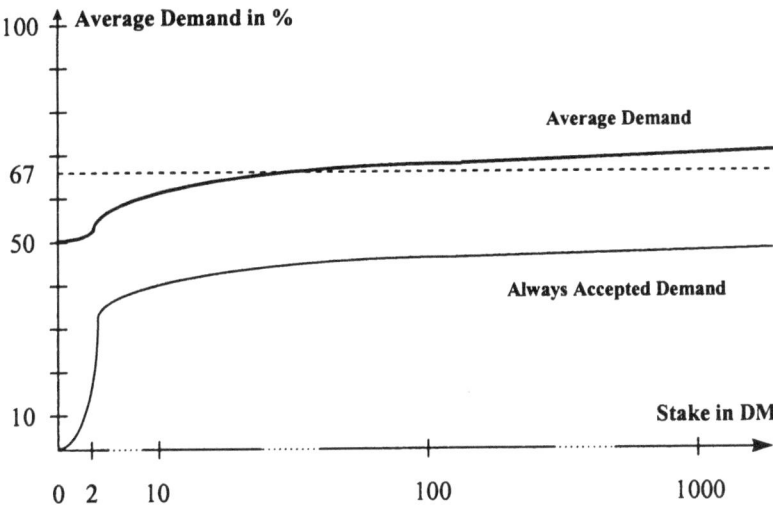

Figure 1: Suggested Average First-Round Ultimatum Demands

rejected demands. Additionally, in case of similar (or corresponding) expectations and according behavior, the demand and accepted demand average is not extremely different due to only few rejections, or no rejections at all. For low stakes, proposers might have perception problems as well. They refer to a 50% split, since they expect acceptance among responders, and do not care about the little money they can gain with a demand of 60% or so. But as stakes become higher and gains perceptible, they become greedier and demand higher shares. They might also anticipate lower rejection frequencies with higher stakes, which is indeed correct as shown in chapter B.II.2. above (Forsythe, Horowitz, Savin, and Sefton 1994, Slonim and Roth 1998). Some more statistical proof is offered by Güth and Tietz (1990), who ran regressions and argue that the best opening move in Ultimatum bargaining is an offer of 40%, since it produces the highest payoff expectation. The relevance of a two third demand was already discussed, identifying a 67% demand as a (sometimes clear-cut) boundary. The average demand appears to be stable among designs, even though some authors argue that a stake of several millions might produce 80 or 90% demands (and no rejections). This might be true, but no statement can be made about average demands. Some demands might be higher, but others might still induce some fairness considerations or fear of rejections. Therefore, it stays unclear how far away from 67% (if at all) the average demand curve will take its course for very high stakes. The existing data does not suggest a significant deviation, which is for example confirmed by the results of Cameron (1995) or Slonim and Roth (1998). Figure 1 might be criticized for its simplicity and underlying assumptions. The two-thirds argument was already discussed, and so were some assumptions. Other points are the fact that Figure 1 is not based on empirical data, but on some unsystematically selected general results and conjectures. Nevertheless, it provides some more insights into the interplay of fairness considerations and fears of rejection.

But how can a norm of fairness depend on monetary stakes? Since for obvious reasons it cannot, the fear of rejection argument provides a far better explanation. Since payoff amounts for responders become more interesting the higher the total stakes, even if their relative share decreases somewhat, proposals are less frequently rejected. Slonim and Roth (1998) show some additional proof. While the rejection frequency in fact decreased the higher the shares, the relative demands did not increase at first. But when more periods were played, proposers raised their relative demands only in the high stakes design. Obviously, the fear of a rejection is lower the higher the stakes, and this also leads to higher demands.

Since fair offers in the Ultimatum game could be seen as purely strategic and maybe be explained by a fear of rejection alone, the results of the Dictator game clearly reject this idea. Of course, subjects take advantage of their improved strategic position as Dictator, which is expressed by far lower offers in the Dictator game. But since these offers are still substantial (usually between 10 and 31%), and even equal splits happen (8 to 23%), a fear of rejection does not serve as the only explanation in Ultimatum games, because this fear definitely does not exist in the Dictator game situation, but considerable offers are nevertheless made. But fairness considerations alone are not sufficient either, as shown by the difference between Dictator and Ultimatum games, see the studies of Suleiman (1996), or Güth and Huck (1997). Proposers take advantage of their strategic power and demand more in the Dictator game. Therefore, the same general norm of fairness is not applicable to both games, or at least not to the same extent. Following this, the hypothetical trade-off between fairness considerations and strategic reasoning seems to take place in both games, but follows different rules or leads at least to other results. The underlying reasons for the differences between Ultimatum and Dictator outcomes have to be highlighted and analyzed to discuss the role that fairness plays in bargaining situations.

C. Fairness and Intrinsic Motivation

The relevance and persistence of fairness for human behavior have been proven by a lot of experimental economists, and were discussed in chapter B. However, fairness seems to depend on several factors, which will be described in the next paragraphs. One of the most interesting factors seems to be intrinsic motivation. This topic and some experimental approaches are addressed first, followed by a general discussion about fairness and the relevance of certain factors for a fairness norm. Following that, the different factors and their possible impact on behavior are put together and briefly discussed.

I. The Concept of Intrinsic Motivation

The discussion about the importance of intrinsic motivation for economic theory has become more intensive recently, documented by the articles of Kreps (1997), Frey and Oberholzer-Gee (1997), Lindbeck (1997) and Frey and Osterloh (1997), to mention just a few. To be able to explain the concept of intrinsic motivation, it is important to clarify its relation to extrinsic motivation. One of the frequently asked questions is whether there is a crowding-out of intrinsic motivation by extrinsic incentives or not. Extrinsic incentives – especially money – are treated as the main motivational force for human behavior in standard economic theories. A rational decision maker who only cares for his own monetary payoff always prefers the alternative with a higher payoff. This might lead to wrong conclusions when people are influenced by their own intrinsic motivation (Frey 1997c). The question is not whether people are motivated by extrinsic incentives rather than by intrinsic motivation. Usually both stimuli are relevant. But an interesting aspect of this situation seems to be the possible crowding-out (or crowding-in) of intrinsic motivation by extrinsic incentives.

In their summary of psychological research in this area, Deci and Ryan (1980) state that the existence of such a crowding-out cannot be denied. Especially children, once paid for doing work in the household, won't work at all without being paid in the future. Frey (1997c) gives a thorough explanation for this phenomenon. Such effects are in harsh contrast to the idea of success-oriented income payments for employees. Frey and Oberholzer-Gee (1997) point out that some companies have therefore already dropped success-oriented payments again. But there may as well be a crowding-in of intrinsic motivation by extrinsic incentives (Frey 1994).

II. Experimental Approaches Towards Intrinsic Motivation

Several experimental implementations towards intrinsic motivation already exist, for example by Falk, Gächter, and Kovacs (1998), who showed that intrinsic motivation is relevant for situations with a significant impact of reciprocity. Bolton and Katok (1998) conducted Dictator game experiments and analyzed the giving behavior in different treatments, finding proof for an extensive crowding-out effect of giving behavior. Frey and Oberholzer-Gee (1997) report on a survey about the acceptance of a nuclear waste repository in a Swiss community. The residents of the community were asked whether they would vote in favor of the repository or not. While more than half of the respondents were willing to accept the repository according to that survey, this changed dramatically when monetary payments were introduced. Faced with regular monetary compensations, only 24% of the residents were willing to accept the repository, indicating that the intrinsic motivation to accept this project was crowded out by the offered extrinsic incentives.

People are intrinsically motivated to perform certain actions without being compensated with money, for example to donor blood, to accept a nuclear waste repository in their community, to work harder as their contract requires, or to give money to people who have less. The social norm of fairness plays an important role in this context. The concept of intrinsic motivation provides an explanation why people act fairly – they are intrinsically motivated to refer to a certain fairness norm in certain contexts. Therewith, intrinsic motivation offers a broader classification of unselfish behavior than fairness. Reciprocity, altruism and other related behavioral observations are most likely triggered by a person's intrinsic motivation. Therefore, these types of behavior should be further analyzed. This is done in the next paragraphs, starting with fairness, and proceeding with the idea of a social norm of fairness.

III. Aspects of Fairness

Some relevant questions for experimental economists concerning fairness might be:

III. Aspects of Fairness

- Are players fair ?
- When are players trying to be fair ?
- Why do players want to be fair ?
- Or are players just pretending to be fair ?
- Or is fairness just a nice phrase ?

In general, the existence of fairness remains unquestioned, and a certain or even decisive relevance of fairness for bargaining situations cannot be denied. However, to give a precise and complete definition of fairness seems to be a rather difficult task, and it should therefore not be attempted to obtain such a definition here. Nevertheless, the meaning of fairness for the observed behavior is extremely important, and some aspects have to be thoroughly considered and finally written down. All of the authors mentioned in chapter B. above refrain from giving a definition of fairness, and only rarely refer to 50:50 splits being perfectly "fair". A more concrete discussion of fairness is usually avoided. This is understandable, since fairness seems to play an obvious and straightforward role in some settings, but appears to be negligible in others. On some occasions, the 50:50 split serves as a good predictor, but is nearly meaningless on others. This can possibly only be explained by an at least somewhat complex approach.

The terminology of fairness also appears to be somewhat confusing. With expressions like cooperation, other-regarding behavior, altruism, solidarity, (good) manners, reciprocity, trust-rewarding, and finally fairness, several different descriptions for some related observed behavioral regularities are in use. Of course, some of these terms reflect specific patterns of behavior for certain types of games. Cooperation is usually used to describe non-egoistic behavior in Prisoner's Dilemma games. But altruism and other-regarding behavior are also common expressions for this pattern. In Public Goods games, the behavior of participants with considerable contributions is often labeled as solidarity. Solidarity is definitely other-regarding, but isn't it altruistic just as well? Roughly speaking, altruism means to care about the well-being of others, and this is again something that is also suggested by solidarity. But it has to be added that altruism sometimes implies caring about the well-being of others more than about the well-being of oneself.

Another dimension is added by the notion of trust-rewarding as well as reciprocity. For the importance of reciprocity, see Fehr and Gächter (1998), for an experimental implementation, see Fehr, Kirchsteiger, and Riedl (1998). Reciprocity means that a fair or cooperative action by the first player is acknowledged (or rewarded) by the second player with a fair or cooperative action in response to the observed move of the first player. This is not solely altruistic, because it might have been triggered by the first player's action alone. But since rational egoistic behavior would still be payoff maximizing, reciprocal behavior is at least partially other-regarding. Of course, reciprocity can also describe a negative response to an initial negative action, i.e. egoistic behavior. For example, rejecting an ambitious proposal in an Ultimatum game could be interpreted towards negative reciprocity. Güth, Ockenfels, and Wendel (1993) show that trust might induce fairness, which would serve as an example for positive reciprocity in that sense. Camerer and Thaler (1995) discuss manners as a possible explanation for the outcomes of bargaining games, and see a close relation to fairness.

This list of discussion arguments could be continued, but it appears to be more productive to focus on concepts that organize or connect the different observations and argumentations. Rabin (1993) uses the term "social goals" to sum up expressions like fairness, altruism, or cooperation. Bolton (1998) considers the concepts of fairness, reciprocity, and altruism, and concludes that their relationship has always been unclear. Bolton (1998) then points out that the main reason for that is the fact that fairness is usually used to describe the behavior in bargaining games, while altruism and reciprocity are common for dilemma games. After stating that the behavioral patterns for both types of games might be closely related, Bolton (1998) illustrates some possible concepts to include the behavior for both game types. While these concepts deal with certain behavior like fairness or cooperation, they do not address them directly, but only cover their impacts, i.e. by formulating a motivation function based on payoff and relative payoff (Bolton and Ockenfels 1999). This approach covers both game types, but does not connect the different terms for behavior listed in this paragraph. The following chapter introduces a new concept to include all of these behavioral expressions, and describes possible applications and shortcomings.

IV. Referring to a Fairness Norm

General thoughts on fairness should start with the heritage of fairness. Where does the role of fairness come from? According to authors like Güth (1995), or Rabin (1993), people refer to a social norm (or social goal) of fairness. If fairness is indeed some kind of social norm, it is obviously not a genetic heritage. But could social norms be genetic heritage? This question does not have to be answered here. But it might be interesting to mention the results of de Waal (1996), who shows that basic patterns of fairness exist in animal societies. Referring to this study, Abbink, Bolton, Sadrieh, and Tang (1998) assume that fairness might serve a deeply rooted biological purpose. This might be true, but it remains unclear which purpose this might be. If fairness as such is not genetic, then it has to be learned. People have to be taught the different applications of a norm of fairness. This view is strongly supported by the observation that children have to be educated to behave according to social norms, since they do not seem to know these norms from the very beginning. Bolton (1997) showed that fairness is evolutionary stable in certain contexts, lending some more support to the idea of fairness as a (stable) social norm. Furthermore, Bolton (1997) argues that fairness is grounded on biological roots. Kreps (1997) explores social norms a little further, and explains why people might adhere to certain norms.

A new behavioral approach would have to take the idea of fairness as a social norm into account, since simple "fair" behavior has to be grounded on something. Authors like Kahnemann, Knetsch, and Thaler (1986a) avoided this debate by calling fair what the majority of their observed population considered to be fair. But this also refers to a somewhat hidden norm of fairness, since the opinion of society was used to measure fairness. This strongly reminds of a social norm of fairness. Following Bolton (1997), splitting fairly is identical with splitting equally, i.e. 50:50. Of course, a large discussion about equity could be started, since equity can be measured by input, output, payoff, or several other variables. Here, the equity problem should also be left aside, but it has to be kept in mind that a 50:50 split might not be the best solution according to a social norm of fairness, and is therefore a dangerous simplification.

Generalizing the notion of fairness to a social norm of fairness implies defining fairness as adjusting behavior to the social norms of society. And the social norms of society are defining the type of behavior that is expected from a person in a certain situation by the majority of that society. By referring to a social norm of fairness, it is possible to reflect all of the behavioral facets listed in C above (solidarity, cooperation, altruism etc.). Therefore, this approach is differing from the simple expression of "being fair", since the latter highly suggests very straightforward behavior, like a 50:50 split. Hoffman, McCabe, Shachat, and Smith (1994) argue that behavior might not be driven by fairness, but by a social concern about what others may think. Again, this just means that a social norm of fairness is applied. If someone expects the majority of the population to refer to such a social norm in a certain context, this norm becomes relevant for his behavior, especially if the outcome is observed by others. Whether he was intrinsically motivated to behave according to this norm or just forced to accept it, does not necessarily change his behavior. Nevertheless, this difference might be important to other research objectives.

Camerer and Thaler (1995) argue that manners might play a key role for bargaining. But manners are also derived from social norms, and therefore show a strong relationship to fairness. If someone shows good manners, you might feel obliged to act politely as well, but if he is rude, you could feel free to be impolite. This is similar to fairness, and good manners might even be called fair, since they clearly express other-regarding behavior just like fairness does.

Other advantages of a concept of a social norm of fairness are easy to find. "Being fair" is mainly a one-sided concept: "A is fair to B". But the social norm of fairness is a global institution, including not only A and B, but also their whole social frame. If both refer to a norm of fairness to plan their behavior, the outcome might be called fair as well. But this only happens in situations where such a norm of fairness exists and dominates other influences. Fairness is not something a person deserves or something that happens to him by chance, but the result of a cognitive process by a group of people or a society or the majority of this society, leading to a clear decision to apply the fairness norm. In some settings, the norm is referred to, in others it is abandoned. Ochs and Roth (1989) avoid concluding that players try to be fair, and claim that it would be enough to suppose that they are trying to take distribution considerations of other players into account. Again, this suggests that a norm of fairness is referred to.

Already the difference between Ultimatum and Dictator games offers suggests that the norm of fairness might play different roles in different settings. Zajac (1995) points out that different institutions have different fairness norms. One could also say that the norm of fairness is applied less consequent in some institutions than in others. While a buy-out of airline passengers in an oversold airplane is accepted, a buy-out of people in a supermarket queue is most likely not tolerated (see Zajac 1995). Kahnemann, Knetsch, and Thaler (1986a) conducted a survey and compared the responses concerning fairness in different situations. They also conclude that judgments of fairness are subject to framing effects, represented by different institutional settings. Hoffman and Spitzer (1985) follow a similar argumentation, which grounds on particular concepts of fairness that are known to each subject. The experimental institutions then trigger certain aspects of that fairness concept, which are then implemented via the subject's decision. Kreps (1997) also suggests that the importance of norms changes depending on circumstances.

Bolton, Katok, and Zwick (1998) describe Dictator behavior as a rules-based decision procedure, partly influenced by fairness considerations. They show that the extent of giving depends on the game frame. For the social norm context, this simply means that the importance of the social norm of fairness differs with the institutional setting, just as discussed above in connection with the work of Zajac (1995), and this results in varying behavior, even for basically the same game. And this also holds for comparisons between Dictator and Ultimatum games. Suleiman (1996) as well as Güth and Huck (1997) show that the fairness norm becomes less important with the decreasing veto power of the responder.

However, the explanation of a varying importance of the fairness norm for different games or settings is ambiguous. A norm of fairness should have some general meaning, or none at all. But there are several ways to solve that problem. The first explanation arises from the setting. Maybe there are types of settings where a fairness norm is usually used, and others where it is simply never applied. This idea is approached in the next chapter. Another reason might be grounded on interpersonal differences, since people might have different ideas of fairness or just different experiences with the fairness norm. Therefore, the applicability of the social norm of fairness in a new setting is at least uncertain, since a lot of different expectations and experiences are suddenly mixed. In their everyday life, individuals learn about norms in general. In new situations, they try to use experiences from similar situations and apply this knowledge, i.e. the use of a fairness norm. The closer a new situation is to a common setting, in which fairness plays a dominant role, the more likely fair behavior should be in this new context. The validity of the fairness norm is questioned the more uncommon that situation is, but also the more knowledge is gained about that unfamiliar setting. When all the rules of a new game are understood, the relevance of the fairness norm for that new situation is once again questioned. This might be the reason behind the surprising results in the Equal Punishment game by Ahlert, Crüger, and Güth (2001), where the relevance of the fairness norm was fragile, since it seemed to explain a lot in the first round, but had nearly no relevance for the repetition. Before deciding about their second round behavior, some or even a substantial group of individuals might have questioned the applicability of the norm of fairness.

Fairness can be called a norm, but not a basic need or desire. Human behavior is driven by needs, and restricted by many social norms. The total impact of all relevant norms for a certain situation might strongly influence the behavior, but will not change the basic needs. But is there a need for fairness? Based on the experimental evidence, the answer is probably yes, but it seems that a norm of fairness mainly serves the purpose of creating a minimum of annoyance, therewith producing maximum payoffs and maybe even efficiency, for example by reducing the number of rejections in Ultimatum games. Kreps (1997) also suggests that people might adhere to a certain norm simply because there might be something desirable about it – in this case less annoyance. Therefore, grounding observations on a social norm of fairness helps to organize different experimental outcomes on a similar basis, and therewith makes comparisons possible. Some more aspects of fairness and annoyance are explained in the next chapter.

V. Relevant Factors for a Social Norm of Fairness

Another difficulty is added by the fact that fair behavior seems to depend on several factors, some of them heavily influenced by certain experimental settings. An anonymity effect does not seem to exist, but gender effects and fixed sacrifice effects were observed. They have certain impacts on fairness, but do not harm the validity of general Ultimatum and Dictator outcomes and the fairness observed there. More crucial effects like the level of competitiveness and the level of social distance might have a more significant impact on human behavior. Therefore, they could be used to generate some explanations for the effects observed in paragraph B.III. above. These two indicators are sometimes closely related, and also connected to another phenomenon called annoyance. The next chapters are aimed to discuss these effects.

1. The Level of Competitiveness

The importance of competitiveness has already been illustrated by Roth, Prasnikar, Okuno-Fujiwara and Zamir (1991), who state that while bargaining is dominated by fairness, a market is not driven by this, and the competitive pressure seems to dominate all fairness concerns. But by analyzing some more of the outcomes listed in chapters B. and B.III. above, another striking fact can be observed. While situations with rather competitive characteristics like markets, auctions, buyer/seller settings, and property right conditions tend to produce equilibrium or close-to-equilibrium results, the more personal or social a game functions, the fairer (and further away from equilibrium in case of an "unfair" equilibrium) is the behavior, as observed in division tasks, personal communication or cooperation settings.

This is caused by the different levels of competitiveness that arise between the players. In a common social and very personal situation like a bargaining setting, competition only plays a minor role. This changes with the nature of the underlying institutions, i.e. markets, or by a pre selection process, i.e. a contest, quiz, or auction, when personal and social factors are dominated by

V. Relevant Factors for a Social Norm of Fairness

other forces. The general design of a game is important, but the instructions add another possibility for variations, since the wording influences the perception of the situation by the participants. As shown by Hoffman, McCabe, Shachat, and Smith (1994), the formulation of a buyer/seller problem generates different outcomes than that of a division problem for the same game. Here, the competitiveness is solely generated by the different experimental frame that is introduced to the participants. Another example is the achievement of positions, either by auctioning or by winning in a contest or quiz. The competition that is experienced in the pre-play phase is also reflected in the behavior of the final game stage.

Who would expect the norm of fairness in a market? A marketplace like a stock exchange does definitely not foster fairness. But Fehr, Kirchler, Weichbold, and Gächter (1998) are able to show that the norm of reciprocity can play a role in certain competitive markets. The close relationship between reciprocity and fairness was already illustrated in chapter C. above. The stock market has norms, restrictions, and laws of its own. This is the social frame of the market – within this frame, competition has no limits, and is neither personal nor socially normed, but anonymous. The more competitive a bargaining situation is formulated, the less fairness will be observed. The closer it is to a real social situation, the more fairness is usually demonstrated by participants. And this might conversely hold true for a market situation – the more personal it becomes, the more likely is fair behavior. Selling stocks to another anonymous (but also private) trader might be less likely to foster fairness than selling a used car to a friendly person. But how about an unfriendly person? The facets of this question are to numerous to discuss here. A final thought should be given to competitiveness in bargaining situations. Schotter, Weiss, and Zapater (1996) create competition among proposers in an Ultimatum game by offering a subsequent second game, which was only available to the best proposer, who is the one who earned the most money in the first round. The results indicate that proposers take the higher level of competitiveness into account by demanding more, and receivers also seem to react to this by accepting low offers more often.

Therefore, a behavioral theory must be well aware of the implications of certain game situations. How do certain instructions (i.e. buyer/seller formulations) influence behavior, since they, for example, import real life experiences, maybe market behavior, into the lab? A hierarchy of situations with differing levels of competitiveness might be helpful, but is certainly not sufficient. An approach based on annoyance is introduced in a later chapter, where the following thoughts about the level of social distance will also play a key role.

2. The Level of Social Distance

In a setting with high competitiveness, i.e. an auction, or high social distance, i.e. a market, annoyance is usually not expressed, either because this is not possible or not desired and therefore suppressed. But in bargaining settings, annoyance is frequently observed when offers are rejected by participants who feel that they have been treated unfairly. In a situation with more social distance like a market, the expression of annoyance is not perceptible. But the more social closeness is added, the more annoyance is likely to be exposed in case of unfairness or other (negative) factors.

Hoffman, McCabe, and Smith (1996b) varied the social distance between proposers and receivers in a Dictator game, and found out that more money is offered the smaller the social distance between participants is kept. Bohnet and Frey (1999a) raise the impact of solidarity by decreasing the social distance between participants in Prisoner's Dilemma and Dictator games. And Brosig, Ockenfels, and Weimann (1999) show that different communication mediums have different impacts on cooperation in a public good game. Thereby, the medium with the lowest social distance, namely personal communication, worked best, and the higher the social distance was kept, the less cooperation was observed. This might also been explained by the different social frames of the respective situations. A social norm of fairness is more likely to play a key role when participants are exposed to situations that are similar to their real life experiences, especially to those with a high relevance of a fairness norm, i.e. with personal contact. In this respect, a low level of social distance and a low level of competitiveness have a similar effect, since both characteristics make it easy for participants to refer to a fairness norm.

3. Annoyance as a Key Factor

The relevance of annoyance for certain behavioral patterns is commonly known, for example as a key factor for the rejections of low offers in Ultimatum games. Receivers are annoyed by the greediness of the proposer and give up their remaining payoff to punish him. It is very likely that proposers are aware of the fact that their unfairness would cause annoyance, and therefore refrain from equilibrium offers. Responder and proposer both refer to the social norm of fairness. And they are more willing to do so in a setting with a low level of social distance or a low level of competition, since the fairness norm is more present in these settings. Therefore, a violation of the fairness norm would cause high annoyance, leading to undesirable outcomes. On the other

V. Relevant Factors for a Social Norm of Fairness   51

hand, with a high level of social distance as well as competitiveness, annoyance might exist, but is less important and also less perceptible, since the fairness norm does not play such a key role here. Again, this is realized by proposers and responders, who behave accordingly, i.e. by equilibrium offers or contributions.

The connection between norms and annoyance is simple – norms should exist to minimize annoyance. Annoyance is caused by several actions, which are for example criminal, rude, or unfair. To prevent people from being annoyed, norms have been established. Behaving according to these norms should avoid any possible annoyance – laws exist to prevent criminality, manners are meant to prevent rudeness, and the social norm of fairness as a special kind of norm should prevent unfairness. Deviating from any of these norms is rejected by the majority of society, and therefore punished. From this perspective, fairness is nothing else but adjusting one's own behavior to the expectations of society. This expected behavior is determined by concrete laws, strict norms or just "soft" manners, and subject to continual enhancements. Of course, it is crucial that these laws are also respected. But if certain norms are not broadly accepted, i.e. income tax laws, this becomes common knowledge, and the useless norm is most likely replaced. A society with meaningful but few norms and people respecting these norms might even be efficient, since the costs caused by deviations and annoyance are minimized. Bolton (1997) argues that social norms persist because they produce efficiency in an evolutionary stable way.

As already mentioned in IV above, there are different fairness principles for different institutions (Zajac 1995). A buying-out of airline passengers in an oversold airplane is accepted, while a buying out of people in a supermarket queue is most likely not tolerated. In the first case, the airline compensates people who are willing to change to a later flight. In the second case, the last person in the queue compensates the first person of the queue with a monetary amount that is accepted by the first person. The value of the saved time for the last person exceeds this amount. All other people are also better of, since this procedure is only used when the first person has more items to pay for than the last person, leading to some saved time for everyone. The first and the last person then change their place in the queue. Both methods should produce Pareto-efficient outcomes. Nevertheless, this is not the decisive factor for the supermarket shopping institution. The social norm of fairness seems to be strong for a queuing procedure, and a market-like intervention is determined to fail, since to much annoyance arises due to the violation of the fairness norm, and this annoyance leads to a strict rejection of the intervention. This is different in the airline travel scenario, where another intervention takes place. People are offered feasible monetary amounts to compensate for their waiting time, while other people take advantage of the resulting available seats. The intervention is not perceived as unfair, and produces only little annoyance. In fact, it might even be welcomed be the majority of travelers, since a further loss of time due to a delay of the take-off is avoided.

The underlying institutions seem to have different fairness principles, and the social norm of fairness is less important in the airplane scenario. The passengers are also less annoyed by that buying-out than supermarket customers. Additionally, Roth, Prasnikar, Okuno-Fujiwara and Zamir (1991) point out that cultural differences in the perception of fairness might exist. They conducted experiments in four different countries, and state that people seem to have different expectations about fair outcomes, and that fairness depends on one's own perception in a respective situation. This might be due to differing experiences with institutions and the meaning of fairness in different cultural environments, and could also depend on varying extents of annoyance that might arise or are expressed differently within other cultural frames. For experimental procedures, this also has a certain impact. People enter the lab with certain experiences, i.e. about the social norm of fairness, manners, annoyance or other expected behavior in certain similar situations. These experiences are impossible to neglect, no matter how neutral the experimental instructions are formulated. Of course, subjects from the same subject pool, i.e. students from one campus, are very likely to have similar experiences.

4. Determinants for a Level of Annoyance

The level of annoyance is influenced by various factors. The most important cause for annoyance are violations of the fairness norm. People are used to real life situations where a great number of norms are minimizing possible annoyance. Confronted with a new situation, they have to get used to the fact that a fairness norm might not have an important impact. In this case, annoyance should perish the more knowledge about the situation is gained, see the Equal Punishment game of Ahlert, Crüger, and Güth (2001) described in B.II.3 above. Another possibility is to conduct a pre-play questionnaire to make participants more familiar with the new situation, and therewith give them the opportunity to form realistic expectations (see Tietz 1992).

Another question is how good experimental instructions form corresponding expectations, since this would reduce the potential for annoyance in experimental settings. The wording is again important, and the use of common words like "market", "buyer", or "seller" induces a certain risk, but also a certain chance to form corresponding expectations, if so desired by the experimenter.

For example, low rejection rates and low annoyance in an Ultimatum game might evolve due to identical expectations about the importance of a fairness norm, which can be reached by explicitly appealing to fairness considerations with words like "division task". The experiment itself might have an influence. Subjects might be annoyed because they were forced to participate, or because they are forced to think, or because they need to make a decision and feel a certain pressure. They might also be annoyed by the random allocation of roles. Especially responders in an Ultimatum game could perceive it as unfair that they could neither chose nor influence the process that was used to determine their own role. And even if they were not annoyed by the allocation of roles, they might be annoyed simply because they have to think about a rejection, since they expect to be confronted with a low offer and a high temptation to reject.

The general validity of fairness as a simple explanation for the observed behavior in Ultimatum and Dictator games appears to be doubtful – fairness does only exist in certain environments where its relevance is assured by contracts, norms or specific social structures. Examples are sports, marriages, or small neighborhoods, while stock markets do function without fairness norms. Therefore, the argumentation has to ground on a social norm of fairness, and on annoyance caused by deviating from this fairness norm, which is again influenced by the level of social distance and the level of competitiveness involved. A theory called ERC (Bolton and Ockenfels 1999) has been developed recently, and aspects of fairness implications are also included in that model, so further comments about the annoyance approach are given after it has been compared to ERC.

VI. Another Implementation of Fairness

The theory of equity, reciprocity, and competition (ERC) by Bolton and Ockenfels (1999) includes fairness in utility via a motivation function, using relative payoff and risk aversion. The idea proposes that people are motivated by their own monetary payoff and their own relative payoff standing. ERC does not claim to be a bargaining theory, even though it organizes the data of a lot of different experiments quiet well, including Ultimatum and Dictator. Further enhancements towards a real bargaining theory are possible. A similar model is introduced by Fehr and Schmidt (1999).

Inequality Aversion somehow does quite well – maybe because it captures the most common thing of humanity – mediocrity. When outcomes are complicated, unpredictable, unknown, extremely hard to calculate or whatever, why not settle for the easiest solution, namely an approach as simple as one, two, three – the same for everyone? But when these outcomes are questioned by laboratory subjects, be it with experience or by applying otherwise gained real world knowledge, strategic aspects receive a higher importance and inequality is accepted and even demanded instead of avoided, for example in the buying-out situations of Zajac (1995).

Why should people care about the payoff of others? This is the relevant question. If someone refers to a norm of fairness, this indirectly has certain implications about the payoffs of all persons involved. If a person does not refer to a norm of fairness, but focuses on his own relative payoff standing, he expresses nothing less than pure envy. This envy might cause annoyance, but this kind of annoyance is not important enough to be suppressed by norms, i.e. there is no society with norms to avoid envy. Of course, in a society with a high relevance of a norm of fairness, envy is avoided automatically. But this norm does not target envy in the first place, but annoyance. And while envy is a somewhat unjustified emotional state, "real" annoyance is caused by violations of relevant norms as described in paragraph C.V.3. above.

Fairness has to be envy-free, and behavioral models should not be envy oriented, but rather annoyance oriented. The only relevant question raised by a division problem is whether a person's share is according to his "rights" or expectations, since this would be perceived as fair. Not important is how much others receive, as long as their share does not violate relevant norms, i.e. of fairness. A comparison with others is not envy-free, and therefore not exclusively motivated by considerations of fairness. This has to be kept in mind to avoid confusion that could possibly be caused by mixing different motivational aspects.

The relevance of concepts like ERC is also questioned by the observations of Güth and van Damme (1998) as well as Selten and Ockenfels (1998), since the frequent unfair or unequal treatment of the third player seems to reject the importance of inequality aversion and the way fairness is modeled here (see also the response of Bolton and Ockenfels 1998). This raises once again the question about the importance of fairness or egalitarian preferences and tackles the approach of inequality aversion (or even hardens the idea or thought of a completely different explanation). Do Ultimatum proposers only offer positive amounts because they fear a rejection? And is this actually smart because receivers do indeed reject since they are frequently annoyed, and this annoyance causes a strong spontaneous reaction, i.e. a rejection? Then maybe, after all, a relatively simple explanation for the behavior in Ultimatum games might be formulated. Human beings simply have a strong tendency to show how annoyed they are. Therefore, they are very likely to use a rejection option if there is one available. Furthermore, the data collected up to date suggests that inexperienced receivers are more annoyed, since they reject more often, and even though rejections don't vanish completely, participants seem to adjust to the impacts of their situation with experience. Nevertheless, the first encounter with such an unbalanced situation seems to annoy people far more than they might even admit to themselves. The two games to be explained later, FTP and RAP, vary the existence of such a rejection option. This option can be chosen or neglected and even traded. Therewith, the two games could be used to explore and manifest this explanation a little further. At last, it would be far more than just marginal to find out that people are driven by (spontaneous) annoyance rather than by a pure natural envy, as somehow implied in the idea of inequality aversion in a person's own disfavor.

Güth (1995) refuses the idea of enhancing utility functions by fairness, altruism or envy. He states that this "neoclassical repair shop" only shifts the problem to another level. Of course, this argument could also be used against annoyance. But annoyance is not meant to be seen as just a single behavioral aspect to be added to an absolute payoff variable (i.e. as some kind of human emotion), but rather as an expression for the whole social context of norms and regulations, incorporated to promote efficiency and minimize annoyance at the same time. And this is just impossible to include in a utility function.

VII. Putting the Factors Together

Not a complete theory is developed here, but an outline of important influences as explained in V above, including the social norm of fairness, the level of annoyance, the level of competitiveness and the level of social distance. A low level of social distance and a low level of competitiveness are very likely to produce little annoyance, since the relevance of a social norm of fairness is obvious and easy to justify in this case. The annoyance approach leads towards a descriptive theory, but is not a bargaining or behavioral theory yet.

According to Hoffman, McCabe, Shachat, and Smith (1994), the tendency towards equal splits can be described as fair. More important, they point out that this just names the observed phenomenon, but fails to explain it. A theory like ERC is based on payoff equality, and reaches remarkable results with this construction. But the underlying motivational forces like fairness norms are not explained. The annoyance approach is different, since the social norm of fairness is taken into account, and the interaction of such a norm with other institutional settings is considered. The resulting level of annoyance serves as a key factor for the determination of behavior.

It has to be mentioned that the annoyance approach takes the norm of fairness into account, but not as direct as ERC, i.e. via equal payoff considerations. Deviations from the fairness norm are only important when a certain level of annoyance is reached by such a deviation. A violation of the fairness norm most likely produces annoyance, but this could be compensated by other effects or also have been caused by other influences, i.e. institutional settings like anonymity or random role allocation. In general, norms are meant to minimize annoyance, and a deviation therefore most likely leads to consequences represented by sanctions, i.e. a rejection.

Figure 2 shows important influences on human behavior and how they interact. The underlying simplifications are strong, but this provides a clear view of the main factors. Not included are expectations as well as interpersonal, intertemporal or intercultural differences. The whole decision situation is surrounded by a frame. This frame might consist of certain laws or norms, but it could also represent a situation without any such restrictions like a market. The institution of a market also has unique characteristics, like the number of competitors or consumers, thereby providing a frame as well. The frame determines the extrinsic motivation and the emotional impact in this specific situation. In Figure 2, such influences are represented by black arrows. The most common example for an extrinsic motivation are monetary incentives. The emotional impact is not only influenced by the frame, but also by the extrinsic motivation and the behavior of others. The extrinsic motivation and the emotional impact of this decision situation have a direct effect on the behavior. But an important third influence has to be considered, which is the intrinsic motivation of the respective individual. This intrinsic motivation has to be seen in relation to the frame, since a certain frame might help to intrinsically motivate someone to refer to the social norm of fairness and therefore, for example, offer an equal split if this is possible within the frame. The concept of intrinsic motivation was discussed in chapter C.I.

In Figure 2, the level of social distance as well as the level of competitiveness is an important characteristic of the frame. The possibly resulting annoyance is captured in the emotional impact. The possibilities for individual behavior are restricted by the frame, and vary strongly between different settings. A situation with the existence of a social norm of fairness should have different influences on human behavior than a situation with the dominance of other rules.

To observe any effects in an experimental setting, the existence of intrinsic motivation and fairness is of course required and therefore has to be proved. In this context, two simple games are developed in the following chapters, which are then implemented by means of an experimental design. The results or some of their aspects can be interpreted to support the existence of intrinsic motivation and of a crowding-out effect. The magnitude or level of a crowding-out can also be determined. But before that, another concept called freedom of choice is introduced.

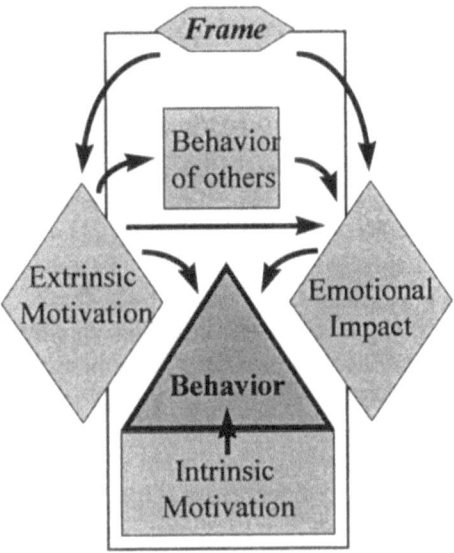

Figure 2: Determinants of Individual Behavior

D. Freedom of Choice

In this chapter, the basic concept of freedom of choice is illustrated and explained. This concept has been introduced by Sen (1988), and offers an interesting field for experimental economists, since no experimental research has yet been conducted. Sen states at the end of his 1988 article about freedom of choice: "The foundational importance of freedom may well be the most far-reaching substantive problem neglected in standard economics". This insight lead to new work in the field of individual freedom, see for example Lindbeck (1988). Other authors concentrate on the handling of axiomatic approaches and the ranking of opportunity sets, e.g. Pattanaik and Xu (1990), Gaertner (1990), or Ahlert and Crüger (1999). The most important thoughts are captured by the following chapters.

I. The Basic Concept

The individual welfare that arises from having the freedom to choose from a given set of alternatives has been analyzed in several studies. According to Sen (1988) and Lindbeck (1988), the act of choosing itself has a certain value (or utility). Additionally, rarely or even never chosen alternatives also have a small, but positive value (or utility). An example would be standard consumer theory. Each household has a consumption optimum. Pattanaik and Xu (1990) point out that providing only this optimum without offering any other consumption bundles reduces the freedom of choice of the household. Another example deals with smoking. If someone neither smokes nor plans to smoke in the future, he might still object against a general ban on smoking, simply because he attaches some value to the freedom to smoke. Sen (1988) describes the concept of freedom of choice as follows.

1. Instrumental and Intrinsic Importance

Freedom has an instrumental importance, since it supports individuals to achieve whatever they consider to be worthwhile. For example, by choosing one bundle of commodities rather than another, an individual is able to increase his or her well-being. Nevertheless, the fact that freedom has an instrumental relevance should not lead to the conclusion that freedom has no intrinsic value in itself. For example, in standard consumer theory the budget set consists of all commodity bundles the consumer can afford. While this budget set might be viewed as a representation of a certain amount of freedom for the consumer, standard economic theory would instead solely concentrate on the value (or utility) of the best available alternative, which is of course the one that is picked by the consumer. Following that, the impact of freedom would not matter as long as the consumer is able to choose his optimal commodity bundle. In the extreme case of no choice with only that best bundle available, the consumer would still be judged as having been treated just like in the situation with choice. This appears to be plausible as long as no intrinsic value is granted to having choices as such. Sen (1988, p. 272) points out that the analysis of freedom should take three categories into account, which he calls "(1) the assessment of the extent of freedom as such, (2) the importance of freedom to individual well-being, and (3) the relevance of freedom in the assessment of the social good and the rightness of actions". These three aspects all have to be considered when arguing about freedom. To be able to show the existence of an intrinsic value of freedom, Sen (1988) continues his characterization of freedom by differing between positive and negative freedom.

2. Negative and Positive Freedom

The positive view of freedom centers on the possibilities and actions that a person can actively choose. The negative approach concentrates on the absence of certain restraints on someone's freedom, which might for example be imposed by the government or other individuals. In economic theory, this negative view has been favored. Sen (1988) admits that both terms might correspond, but rejects the dominance of the negative perspective. He points out that positive action might be necessary to defend negative freedom when under attack. Following this, it appears to be essential that negative freedoms are incorporated together with a disvalue of their violations.

In addition to that, the purely negative view is criticized by Sen (1988) since it leaves cruel situations or hardships like famines out of account. Even in a system with perfect negative freedom, famines happen despite food availability, because certain groups of the respective society are faced with useless rights that do not prevent hunger. Following this, there seems to be no justification to concentrate on negative freedom alone, and also a certain need to focus on overall freedoms. This includes the possibility of a person to do something or to be something, for example well nourished. Of course, negative freedom still has to be acknowledged on its own and also for what it contributes to positive freedoms, since the absence of restrictions enhances the possibilities of an individual to take advantage of his positive freedom. A description of the aspects of positive freedom is given in the next paragraphs.

3. Alternative Spaces, Functionings, and Capabilities

The characterization of positive freedom is somewhat more complex. In standard economic theory, real income and commodity bundles serve as indicators for individual well-being. From this point of view, freedom can be measured by means of alternative commodity bundles from which the individual is able to choose. The problem of this approach are the strong interpersonal differences regarding the transformation of income into a possibility to do or to be whatever is desired. For example, there are persons who might need more food due to a larger body size or higher calorie usage rate. Therefore, if positive freedom should be measured by the possibility of an individual to achieve certain functionings, the field of commodities has to be enhanced. A possible approach is introduced by Rawls (1971) with the idea of primary goods. These primary goods, like the basic liberties or the freedom of movement, enable individuals to choose the kind of life they want to live. Under this argumentation, the index of primary goods serves as a profound measurement for freedom. This leads away from seeing freedom in purely negative terms, and also leads beyond concentrating on commodity ownership alone. However, this approach has another shortcoming, since it does not take the interpersonal differences into account, which arise from more needs or different abilities to take advantage of primary goods to receive whatever is desired. These differences in the capability to function are not reflected by the index of primary goods. Sen (1988) therefore states that freedom should be measured in terms of alternative bundles of functionings from which a person can choose. This idea is illustrated by an example in the next chapter.

4. The Famine Example

The freedom to lead a long life is extremely valuable to the vast majority of human beings. Therefore, the length of life expectancy may serve as an indicator of positive freedom. The ability to avoid early death is nothing else but the positive freedom to achieve a much desired functioning, and it therewith represents a mighty capability.

With respect to that, Sen (1988) points out that positive freedom has to be seen in the right space, which are functionings and capabilities, and not commodities or income. A good income might of course be meaningful to reach sufficient nutrition, but if a person is not able to achieve the functioning to live long despite his income, the shortcoming of the income approach is quite obvious. The impact of that difference is plain to see in Sen's (1988) example, which compares five developing countries with regards to income and life expectancy (see Figure 3).

Country	GNP per head in 1984 (US Dollar)	Life expectancy at birth in 1984 (years)
South Africa	2.340	54
Mexico	2.040	66
Brazil	1.720	64
Sri Lanka	360	70
China	310	69

Source: Sen (1988), p. 280

Figure 3: Income and Life Expectancy in Five Developing Countries

Figure 3 includes the gross national product (GNP) per head as well as the life expectancy at birth for 1984. Five developing countries are considered and sorted by the GNP per head. Even though countries like China and Sri Lanka show a comparably low GNP per head, these countries were nevertheless able to provide a significantly higher life expectancy than South Africa, even though the GNP in South Africa is more than seven times higher than in China and more than six times higher than in Sri Lanka.

This contrast illustrates the shortcomings of the approach of using income as a measurement for freedom. The advantages of measuring freedom in terms of alternative bundles of functionings from which the individual can choose has been demonstrated by the above example. The freedom to choose the functioning to lead a long life is extremely valuable, but not covered by an evaluation based on pure income alone.

On the other hand, the usage of income as a measurement for the well-being of an individual is very common in standard economics. It is implied that this income leads to well-being, since an individual is supposed to have enough freedom to spend his income as he likes to maximize his well-being. And usually, an individual with an average income is able to purchase food. But during famines, there is either not enough food available or the market system does not function properly. And if the government does not intervene, people are left starving. Therefore, the difference between the income and the functioning approach becomes especially visible in such an extreme situation.

Sen's example can be criticized for its simplicity. The GNP does not necessarily reflect the income standard of the poorest citizens. It might be that the greatest part of the GNP is earned by a small fraction of the total population, while the majority of citizens does not participate in that wealth. In this case, the government has to reallocate income by a distribution system or the poor majority has a high risk of dying at an early age due to their poverty. This might be the case in South Africa, where a white majority used to own a great part of the land. With respect to that, the GNP alone is not a good indicator. The real income of the poorest group of the society would be much more accurate, since this group is most likely to suffer death from famines.

This leads to the role of the government in supporting freedom. Sen (1988) states that the higher life expectancy in China and Sri Lanka has been reached by public policies, namely the distribution of food, medical provision and health care. By means of that, these interventionist public policies enhance positive freedoms. They support the ability of individuals to reach important functionings, i.e. to consume food and to lead a long life. In this context, Sen is also able to enlighten the instrumental role of freedom by demonstrating that the instrumental use of political freedom, civil liberties, and especially journalistic liberty and open political opposition has also contributed to avoid famines in India, since they have all served as an early warning system for the government. In contrast to that, China has suffered from a lack of these liberties and experienced a disastrous famine from 1958 to 1961, which might have been avoided if a free press had warned the public and the government, and also put some pressure on politicians as well.

Assar Lindbeck (1988) analyzes the interdependencies between public interventions and individual freedoms even further, and concludes that public policies should not be evaluated based solely on the commodity bundles that can be consumed by the individuals. Other important points to consider include the difficulty of the individual to change his own economic situation, frustration due to restrictions like rationing, a removal of "apparently irrelevant" alternatives, reduced predictability of the outcome of personal choices, and reductions in integrity or privacy. For all of these reasons,

Lindbeck (1988) summarizes that public interventions have a natural limit, beyond which the freedom-reducing effects overweight the benefits. Following that, positive freedom can be enforced by the government, but a substantial part of it should stay with the individuals and must not be influenced by public policies. Returning to Sen (1988), it should be kept in mind that choosing itself might be an important functioning, giving evidence to the thought that freedom has a substantive value for several areas of economic theory and policy making. The following chapter shows some additional work in the area of freedom of choice, namely the axiomatic modeling.

II. Axiomatic Modeling of Freedom of Choice

To be able to evaluate the extent of freedom of certain situations, a clearly defined axiomatic modeling might be of assistance. Pattanaik and Xu (1990) demonstrate a possibility to rank opportunity sets which are available to individuals, in this case to households. They consider freedom of choice by implementing three axioms that take the degree of freedom for the choice maker into account. The result is that the degree of freedom of choice can be based on the pure number of available options – the more the better. This might appear to be somewhat trivial, but their approach served as a solid basis for research to come. Gaertner (1990) enhances this approach by considering information gathering costs, and points out that a smaller set of options can be better if the collection and processing of information becomes too costly.

Another axiomatic approach by Ahlert and Crüger (1999) is aimed at narrowing down a concrete value for freedom of choice. By introducing a welfare change function, it is possible to measure a welfare loss caused by the removal of an alternative. More specifically, the value of freedom of choice is estimated by modeling an upper and a lower bound for this welfare loss. Another possibility to assess the value of freedom of choice is by experimental evidence. The next chapter illustrates this line of investigation.

III. Modeling Freedom of Choice with a Simple Game

One might ask how much value an unchosen alternative may have, since it can not be priced or otherwise directly measured. This question could be experimentally examined by letting people pay to retain alternatives that are not useful at that moment. The players receive a bonus if they exclude certain alternatives before they actually enter the final decision process. The objective of the analysis is to investigate whether individuals prefer to have some freedom of choice or to have no choice dependent on the size of the monetary payoffs. One of the results could be a certain payoff or bonus-amount, which is just the sum that makes the individual either chose the situation without choice or makes him indifferent between a situation including the alternative and the situation without this alternative. Such results would suggest the existence of a positive value for freedom of choice. An experimental implementation based on the ideas explained in this paragraph is illustrated in chapter E. The experimental results, following after that, seem to prove the existence of such an indifference or bonus amount.

IV. A Summary on Freedom of Choice

The discussion about freedom of choice is sometimes pretty philosophical, and tends to use terminology from existing theories that deal with different aspects of freedom. Some of these theories appear to be contradictory about the role of freedom, but as far as freedom of choice is concerned, they simply rely on different assumptions or use certain terms differently. Sen (1988) himself admits that the wording is sometimes the problem with diverging theories or different interpretations. By introducing and clearly defining the different meanings of positive and negative freedom as well as the functionings approach, Sen is able to clarify the vagueness of the term "freedom" and demonstrates its importance for economic theory. Some shortcomings of Sen's argumentation, like the income problem of the famine example, have already been criticized above. But altogether, the importance of freedom of choice, and the existence of both its instrumental and intrinsic value, remains unquestioned. However, there appears to be a measurement problem. Sen has provided some qualitative aspects, but no quantitative approach, which is usually so important to economics.

At this point, the axiomatic approaches might come into play. The insight of Pattanaik and Xu (1990) to simply count alternatives might sound trivial, but the importance of the existence of alternatives seems to be hard to reject in the light of the discussion about the importance of freedom of choice. By following the idea of isolating a value for the welfare loss caused by an excluded alternative, the theory of freedom of choice could receive some additional support and also valuable possibilities for enhancement, see Ahlert and Crüger (1999). To avoid confusion, it should be mentioned that the concept of freedom of choice is analyzed independently from considerations of fairness or intrinsic motivation. Before this chapter on freedom of choice, the phenomena of fairness and intrinsic motivation were discussed in chapter C. Freedom of Choice should not be seen as a rivaling theory, but rather as another aspect or simply another point of view, focusing on rather technical aspects of the choice situation. The idea of fairness remains untouched. In the course of this study, both concepts will be validated against the experimental results. The approach for an experimental investigation and also the two underlying games are illustrated in the next chapters.

E. The Two Games and Their Experimental Realization

The concepts of fairness and intrinsic motivation as well as freedom of choice were already discussed in chapters C. and D. Now, some experimental evidence for those concepts should be obtained, either about their general validity or at least towards some new insights into the underlying reasons or forces. An experimental approach towards freedom of choice has not been realized before. It appears to be difficult to proof the existence of freedom of choice, since it does not have or produce a tangible value per se, e.g. a market price. The nature of freedom of choice induces that it is impossible to measure or judge it in general. But according to Roth (1995), it appears to be a valuable task to spend some effort on learning how to model as games those situations that someone wants to study. Therefore, the game "Freedom to Punish (FTP)" was developed not only to analyze a completely new model, but also to fulfill the task of modeling an existing concept. The game features the new idea of a salable right to punish.

Just like in Ultimatum and Dictator games, a proposer decides about the distribution of a fixed monetary amount. But before that, the receiver has to choose whether he wants to play the game with a rejection option or without. This decision is private information for the receiver. For excluding his rejection option, the receiver obtains a monetary bonus in addition to his payoff. This means that the receiver has the possibility to sell his right to punish before the actual bargaining or division task starts by means of a secret decision for or against his own veto power. It is then possible to explore under which circumstances the receivers sell their veto power, for example by using various monetary incentives, and also no incentives at all. By comparing the decisions in different designs, some insights into the value of such secret veto power could be gained, and some aspects of freedom of choice could be observed. By selling their veto power, receivers give up some freedom of choice. The value of this monetary useless alternative of keeping the veto power could be discovered.

From another point of view, the receiver has to make a general decision about the shape of the bargaining process. It might provide new insights into bargaining behavior when this new stage of the bargaining process is analyzed. Usually, the strategic power of the proposer is achieved by chance, when the participants are seated and receive their roles from the experimenter. But other

E. The Two Games and Their Experimental Realization

methods have also been used, for example allocating roles based on an observable coin toss, on a quiz or contest, or on an auction. In the new FTP game, the receiver has the possibility to influence the rules of the bargaining process, even though his decision remains unknown to the proposer. Of course, the receiver still has to adhere to his randomly achieved role, but might feel more involved in the game than he would as a normal Dictator or Ultimatum game receiver, and therefore possible sources of annoyance are diminished. By selling their veto power, some receivers might also try to avoid having to feel annoyed, since they exclude the necessity to make a decision in the last step of the game about a rejection of unfair offers.

The comparative model of Bolton (1991) induces that receivers are willing to pay a cost to obtain a division that they consider fair. This cost is represented by the substantial shares that are given up by rejecting (unfair) offers. Since Ultimatum responders are often willing to pay this price in the final step of the Ultimatum game, it would be interesting to know whether they are willing to pay some of this price in advance in the FTP game. This price is the bonus amount that is turned down by choosing veto power. Another way to explore this would be by demanding a certain price for the veto power, instead of paying a bonus for exclusions of the veto power. To keep the bonus and the price design comparable, the proposers can be endowed with exactly the amount of that price as a kind of show-up fee.

In the FTP game, the decision about the veto power is private information for the receiver. Of course, it would be very interesting to compare this situation with another setting in which the proposer receives the information about the rules of the game, i.e. whether the receiver has a veto power or not. This is the case in the game "Right and Choice to Punish (RAP)". The veto power decision of the receiver is visible for the proposer, who then chooses one of only two available distributions of the cake. According to this, the RAP game has a cardinal character, a simplicity that makes it easier to determine whether a crowding-out is taking place or not. By varying the available distributions between a fair and greedy combination and a greedy and very greedy combination, the impacts of different monetary incentives on intrinsically motivated fair behavior are observable.

An additional aspect of fairness is added by the fact that the bonus in the RAP game is paid to both players in case the veto power has been sold. This means that a positive action of the receiver, i.e. refraining from his right to punish and therewith giving the proposer the bonus, could produce reciprocal behavior on behalf of the proposer, i.e. choosing the distribution of the two available which best represents a fair outcome. Furthermore, the general impact of a participation of the receiver in the determination of the rules of the bargaining process could be explored. A decision for a veto power might signal a high willingness to reject unfair proposals, and therefore lead to a high percentage of fair distributions.

Both games offer the opportunity to gain further insights into the role of fairness in bargaining. Especially the involvement of the mechanism design option for the receiver is supposed to be of interest. Other designated research objectives are the test for the relevance of a two third demand by proposers, the acceptance behavior in comparison to standard Ultimatum games, and maybe other comparisons. It is extremely important to mention that the two new games are neither Ultimatum nor Dictator games, but new and totally different games with unique dynamics and strategic possibilities. Nevertheless, comparisons with existing studies are possible with respect to these differences. Both games are described in more detail in the following chapters.

I. Freedom to Punish

The game Freedom to Punish (FTP) received this name because the receiver has the possibility or freedom to chose a situation with a punishment option or to abandon this option and continue without being able to punish in the future. This could also be called veto power, since the punishment is actually the rejection of a proposed distribution of a certain monetary amount. The FTP game is a two player, three stage, anonymous and noncooperative bargaining game. Even though it has certain similarities with the Ultimatum and the Dictator game, the FTP game represents a new and considerably different situation.

1. The Structure of the Game

A cake C has to be distributed between two players, receiver R and proposer P. This distribution takes place in three steps.

Step 1: The receiver R chooses whether he wants to have veto power (VP) or not (NV). In the situation without veto power, R (only) receives a bonus δ in addition to his payoff. The proposer P is not informed about R's decision.

Step 2: P makes an offer about the distribution of a cake C. A distribution yields a payoff of y for him (P), and therewith a payoff of C-y for R.

Step 3: R can only reject the proposed distribution if he had chosen veto power in step 1. In this case, he can either accept the offer (take – T) or reject it (leave – L). If he rejects, the payoffs of both players are zero. If he accepts, the proposed distribution is realized. If R had abandoned the veto power in step 1, he (R) receives a bonus δ in addition to his payoff, but he does not have to make a further decision in step 3.

The FTP game in extensive form with a constant bonus δ is shown in Figure 4. Another possibility is a proportional bonus, but these issues will be discussed later.

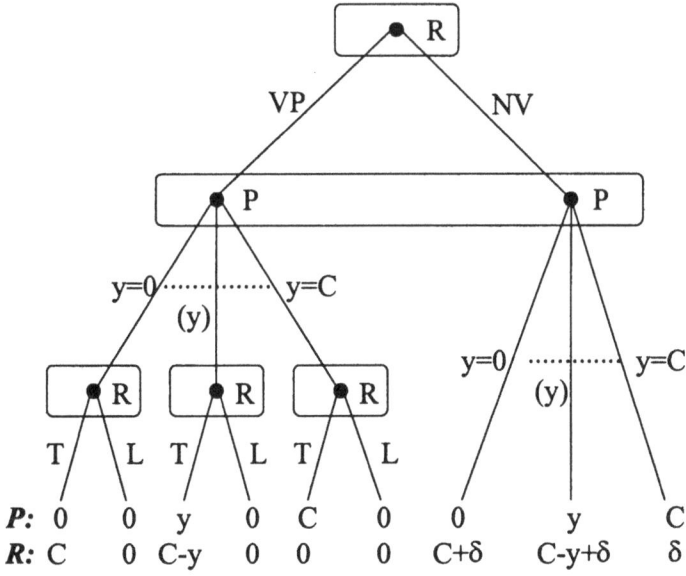

Figure 4: Game-Tree with Parameters for the FTP Game

2. The Game Theoretic Solution of the FTP Game

All of the parameters of the game are positive, $C > y > 0$, $\delta \geq 0$. This also means that the pie-share for the receiver (C-y) is always positive. Figure 5 shows the simplified payoff matrix for veto power and without veto power and for a demand of $y = C$ and $y = C - \varepsilon$, whereas ε is the smallest possible unit, i.e. DM 0,01. Rational behavior is easy to determine in this game.

	P	
R	VP	NV
$y = C$	C (given T) 0	C $0 + \delta$
$y = C - \varepsilon$	$C - \varepsilon$ ε	$C - \varepsilon$ $\varepsilon + \delta$

→→→→→→→→

Figure 5: Payoffs for Demands of C and C - ε

In stage 1, R has to decide whether he wants veto power or not. In any case, the receiver can gain more without veto power due to the bonus δ, as can be seen in Figure 5. NV always yields a higher payoff. Since R cannot signal and threat with VP, he can only behave strategically, play NV and gain the bonus therewith. The only case in which his gains of choosing NV are zero is a zero offer and a proportional bonus, since the bonus is then also zero. But he is just indifferent in this single case, and therefore his best choice remains NV, which is indicated by the black arrows below Figure 5.

In stage 2, the proposer demands as much as possible. Since the NV-choice can be anticipated by P, he demands the whole cake $y = C$.

In stage 3, the receiver would only act in the VP scenario. Just like the normal Ultimatum game solution, R always accepts offers bigger than zero. He is indifferent between taking or leaving an offer of zero. A possible rejection of a demand of $y = C$ in case of veto power should be considered shortly. By rejecting C, R forces P to offer at least ε. This is all R can expect to gain. But as long as $\delta > \varepsilon$, R has no interest in choosing VP, and therefore threatening with L is useless. For the special case of no bonus, $\varepsilon > \delta = 0$, the game theoretic solution changes. The responder now chooses VP, since there is no more bonus to gain by playing NV. The proposer then has to offer ε to avoid a rejection. In case of $\delta = \varepsilon$, the receiver chooses NV due to the safe payoff of δ.

The complete equilibrium strategy for player P is S_P^*:

$$S_P^* = \begin{cases} y = C & \text{for } \delta > 0 \\ y = C - \varepsilon & \text{for } \delta = 0 \end{cases}$$

The complete equilibrium strategy for player R is S_R^*:

$$S_R^* = \begin{cases} NV, \begin{cases} T \text{ for } y < C \\ L \text{ for } y = C \end{cases} & \text{for } \delta > 0 \\ VP, \begin{cases} T \text{ for } y < C \\ L \text{ for } y = C \end{cases} & \text{for } \delta = 0 \end{cases}$$

The complete equilibrium strategy for both players is S^*:

$$S^* = \begin{cases} NV, y = C, \begin{cases} T \text{ for } y < C \\ L \text{ for } y = C \end{cases} & \text{for } \delta > 0 \\ VP, y = C - \varepsilon, \begin{cases} T \text{ for } y < C \\ L \text{ for } y = C \end{cases} & \text{for } \delta = 0 \end{cases}$$

II. Right and Choice to Punish

The game Right and Choice to Punish (RAP) received this complicated name due to the unique combination of alternatives that are available to the responder. First, he can chose between a situation with a Right to Punish and a situation without this right, and then later he can only decide to punish if he is in the situation with this punishment option. This option is basically a veto power, since the punishment is actually the rejection of a proposed distribution of a certain monetary amount. The RAP game is a two player, three stage, anonymous and noncooperative cardinal bargaining game. Even though it might be somewhat related to cardinal Ultimatum and Dictator games, the RAP game offers a new and distinctively different situation. The major differences to the FTP game are the cardinal character of RAP, the missing private information of the veto power decision and the handling of the bonus, which is kept solely proportional in RAP and paid out to both players. A more detailed comparison is conducted in a following chapter.

1. The Structure of the Game

A cake C has to be distributed between two players, receiver R and proposer P. This distribution takes place in three steps.

- Step 1: The receiver R chooses whether he wants to have veto power (VP) or not (NV). In the situation without veto power, both players receive a bonus δ in addition to their payoffs. The proposer P is informed about R's decision.

- Step 2: The proposer chooses between two given distributions of C. One of these distributions yields a payoff of y for him (P), and the other a payoff of Y, whereas $Y = y + \Delta y \leq C$, $\Delta y > 0$ and therewith $Y > y$. Therefore the payoffs of R are either (C - y) or (C - Y), respectively. R is informed about P's decision.

- Step 3: R can only reject the proposed distribution if he had chosen veto power in step 1. In this case, he can either accept the offer (take – T) or reject it (leave – L). If he rejects, the payoffs of both players are zero. If he accepts, the proposed distribution is realized. If R had abandoned the veto power in step 1, both players receive a bonus δ in addition to their payoffs, but R does not have to make a further decision in step 3.

Refer to Figure 6 for a formal presentation in extensive form.

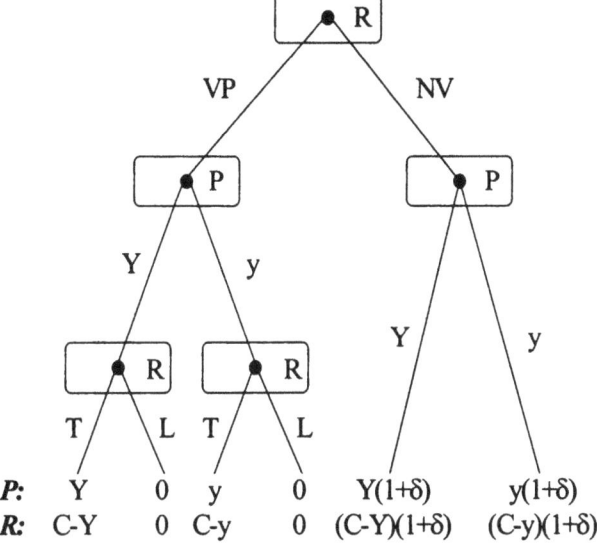

Figure 6: Game-Tree with Parameters for the RAP Game

2. The Game Theoretic Solution of the RAP Game

All of the parameters of the game are positive, $C > Y > y > 0$, $\delta > 0$ (a bonus of zero is excluded for the RAP game). This also means that the pie-share for the receiver (C-Y or C-y) is always positive. Therefore a game-theoretic solution can be determined rather easily. In the last step, R has to choose T, because this always generates a higher payoff for him than L. In step 2, P therefore has to choose Y, since this always yields a higher payoff than y. And in the first step, VP is dominated by NV, because R only receives a bonus by choosing NV.

The complete equilibrium strategy for player P is S_P^*:

$S_P^* = (Y, Y)$.

The complete equilibrium strategy for player R is S_R^*:

$S_R^* = (NV, T, T)$.

The complete equilibrium strategy for both players is S*:

S* = (NV, Y, Y, T, T).

III. Differences and Similarities Between the Two Games

Some factors clearly distinguish between the FTP game and the RAP game.

- The main difference between the two games is caused by the divergent available information. In the FTP game, the proposer is able to observe the veto power decision of the receiver. In contrast to that, this veto power decision becomes public knowledge in the RAP game.
- While the bonus for abandoning the veto power in the FTP game is paid only to the receiver, both players obtain this bonus in the RAP game.
- The RAP game is a cardinal game, since only two allocation choices are available to the proposer. In the FTP game, the proposer can distinguish between all possible allocations, i.e. 1.001 in case of a DM 10 cake.

Even though the two games show these substantial differences, the game theoretic solutions are much alike, since in both settings receivers have to sell their veto power and accept whatever is offered, and proposers should demand as much as possible. The two games have some basic similarities, simply because they are two player, three stage, anonymous and noncooperative bargaining games. The experimental realization for both games is developed in the next chapter.

IV. The Experimental Realization

This chapter provides an overview of the methods that were used for the experimental implementation of the two games. While seven different designs were realized for the Freedom to Punish experiments, which are labeled with alphabetic letters from A to H, the eight designs for the Right and Choice to Punish experiment are classified using Roman numbers from I to IV, and an additional letter in certain cases. The following Figure 7 lists all realized experimental sessions.

IV. The Experimental Realization

No.	Design	Date	Description	Observations	No. of Persons	Sum of Payoffs
1	A	20.10.97	Pilot FTP	20	41	50,75
2	B	30.11.98	FTP 10% Bonus	20	40	194,35
3	C	30.11.98	FTP 0,- Bonus	20	40	160,-
4	D	30.11.98	FTP 0,50 Bonus	20	40	198,-
5	E	30.11.98	FTP 2,- Bonus	20	40	226,-
6	F	30.11.98	FTP 0,50 Price	20	40	185,50
7	G	not played	FTP 1,- Price			
8	H	30.11.98	FTP 2,- Price	20	40	230,-
9	I	29.01.98	RAP FGS	8	16	104,-
10		4.05.98		12	24	206,-
11	II	29.01.98	RAP VGS	9	18	163,-
12		4.05.98		11	22	142,-
13	III	4.05.98	RAP FGH	20	40	391,-
14	IV	4.05.98	RAP VGH	20	40	480,-
15	IU	30.11.98	Ultimatum FGS	10	20	140,-
16	ID	30.11.98	Dictator FGS	10	10	130,20
17	IIU	30.11.98	Ultimatum VGS	10	20	140,-
18	IID	4.05.98	Dictator VGS	5	5	87,15
19		30.11.98		4	4	67,20
			TOTAL:	259	500	3.295,15

Figure 7: Session Overview for Both Games FTP and RAP

The Freedom to Punish game was organized in sessions 1 to 8, using designs A to H. All other designs are based on the game "Right and Choice to Punish", and were realized in sessions 9 to 19. A first pilot experiment based on FTP design A was conducted with students of the Martin-Luther-Universität Halle on October 20th, 1997. The 41 participants were all advanced students of economics or business administration. Designs DI, DII, DIII and DIV for the game RAP were realized on January 29th, 1998 and May 4th, 1998, with 165 participants. Twenty pairs of players were confronted with each of the four designs, which produced 160 strategies. On the same occasion, design IID was played with 5 people, using 5 dummies as Dictator game receivers. Designs B,

C, D, E, F, H were played on November 30th, 1998, as well as designs IU, ID, IIU and IID with 294 participants in total. Due to budget constraints and other reasons, design G was not played in the context of this study. Altogether, 500 participants were acquired for the experiments concerning the FTP and the RAP game.

V. The Experimental Procedure

The sessions were held as classroom experiments during lectures in economics or business administration with students of economics or business administration of the Martin-Luther-Universität Halle. The corresponding lectures of the respective experiment dates were game theory on October 20th, 1997, decision theory on January 29th, 1998, fiscal theory and policy on May 4th, 1998 and introduction into economics on November 30th, 1998.

The seating of the participants was organized with a seating plan, keeping at least one row and several seats unoccupied between participants. Different designs and roles were not mixed, but each group playing the same design was separately seated. The matching for the experiments therefore happened by chance, depending on the seats that were randomly chosen by the participants. The experiments lasted roughly 30 minutes.

For the payoffs, no double blind procedure was used. Instead, a kind of indirect blind payoff procedure took place. The payoffs were not handed out by anyone involved in the experiment, but by neutral administrative personnel. This was also announced to the participants before the experiment. The total payoff for all sessions was DM 3.295,15 (Euro 1.685 or US$ 1.819 at that time), leading to an average personal payoff of DM 6,59. Considering the usual student hour wage of about DM 15, the monetary payoffs should generate an acceptable level of motivation for (rational) behavior. The experiments were financed by Deutsche Forschungsgemeinschaft (48%), Gesellschaft für Experimentelle Wirtschaftsforschung (18%) and other funds (34%).

F. Experimental Design for the FTP Game

The main motivation for the design of the FTP game is to explore under which circumstances the receivers sell their veto power. Therefore, an experimental design has to model different situations, for example by using various monetary incentives or no incentives at all. By comparing the decisions in these different designs, insights into the value of such secret veto power can be gained, and some aspects of freedom of choice could be isolated. Several different bonus types can be used. One possibility is a constant bonus. This kind of bonus is realized by paying a fixed additional sum to the receiver if he has abandoned his veto power. A proportional bonus is the second alternative. The payoff of the receiver is raised by a certain percentage. The outcomes for these two bonus types then could be compared, but they should also be compared to a control design with a bonus of zero.

Another design enhancement distinguishes between a constant bonus and a constant price. Designs with a constant bonus and designs with a constant price are identical concerning the game and the payoffs, because the receiver achieves an additional monetary amount for abandoning his veto power. The only difference is the story that is told to the participants. In the constant bonus design, the bonus is paid by the experimenter to the receiver if the veto power has been abandoned. In the constant price design, the receiver is endowed with a bonus amount from the very beginning of the experiment. To obtain veto power, the receiver has to pay that bonus amount to the experimenter. In other words, the receiver can keep the bonus by refraining from veto power, but has to pay it as a price to receive veto power. Again, the price and bonus designs have to be compared, and both should also be compared with the zero bonus design.

The third possibility is to implement different bonus sizes. A high and a low bonus should be realized, and their final value has to be adjusted to the size of the cake that is distributed. Again, the outcomes for different bonus sizes can be compared, and also have to be compared with the zero bonus design. Now, the different designs are explained in more detail. After that, the hypotheses for the FTP game are developed, as well as some additional design alternatives.

F. Experimental Design for the FTP Game

I. Design Approach for the Experiment

Figure 8 shows the realized designs for the FTP game, labeled A to H. Design G was never played, since the results for designs F and H appeared to be sufficient.

No	Design	Date	Description	Observations	No. of Persons	Sum of Payoffs
1	A	20.10.97	Pilot FTP	20	41	50,75
2	B	30.11.98	FTP 10 % Bonus	20	40	194,35
3	C	30.11.98	FTP 0,- Bonus	20	40	160,-
4	D	30.11.98	FTP 0,50 Bonus	20	40	198,-
5	E	30.11.98	FTP 2,- Bonus	20	40	226,-
6	F	30.11.98	FTP 0,50 Price	20	40	185,50
7	G	not played	FTP 1,- Price			
8	H	30.11.98	FTP 2,- Price	20	40	230,-
			TOTAL:	140	281	1.244,60

Figure 8: Design Overview for the FTP Game

All designs were realized as classroom experiments, with simultaneous decisions of the receiver (step 1 of the game) and the proposer (step 2). After that, the receiver obtained the resulting information for step 3 of the game, in which he either made his decision about the acceptance or rejection of the demand, or simply read about his payoff in case of no veto power. Only one round of the game was played in each of the designs. This strict one-shot approach was chosen to gain first insights into the dynamics of this new game. Playing several rounds induces influences of learning or path dependence, and therefore has to be done by future research.

I. Design Approach for the Experiment

1. Treatment Variables

The parameters of the game can be varied to observe and isolate certain aspects (changes or inconsistencies) of decision behavior. There are two parameters:

C   The size of the pie is held constant.

δ The bonus can be varied.

For all FTP sessions, the size of the cake is DM 10. The bonus is varied, and this includes the size, the type and the kind of payment. Therefore, the resulting 3 x 2 x 2 design structure is established according to Figure 9. For the proportional bonus type, only a low bonus size was realized, and it was exclusively paid out as bonus. The zero bonus design serves as proportional, constant, bonus payment and price payment design at the same time.

Bonus Size	Type of Bonus	δ paid as Bonus	δ paid as Price
Low: δ = 10%	proportional	Design A and B	-
δ = 0	-	Design C	Design C
Low: δ = 0,50	constant	Design D	Design F
High: δ = 2,00	constant	Design E	Design H

Figure 9: Design Structure for the FTP Game

2. Designs with a Low Proportional Bonus: A and B

Design A served as a pilot study, and the players were paid out with a probability of 0,25. Five of the participating 20 pairs of players received their payoff. Design B was nearly identical, but of course all of the participants were paid here, as well as in the remaining designs. Additionally, the players were asked to answer questions regarding their expectations about the behavior of the other player. This was not done for design A, but for all other designs. A value of 10% was chosen for the low proportional bonus. The game tree for designs A and B is shown in the following Figure 10.

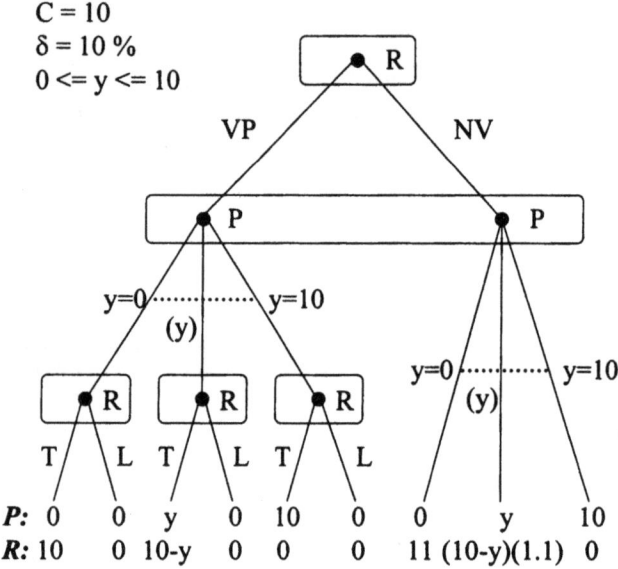

Figure 10: Game-Tree for Designs A and B with a Bonus δ of 10 %

3. The Design Without a Bonus: C

To be able to formulate conclusions about the impact of a bonus, the game was also played with a bonus (or price) of zero. This means that veto power can be obtained without cost. The resulting design C is comparable to all other designs, no matter if bonus payment, price payment, proportional, or constant bonus rules were used. The resulting game tree follows with Figure 11.

I. Design Approach for the Experiment

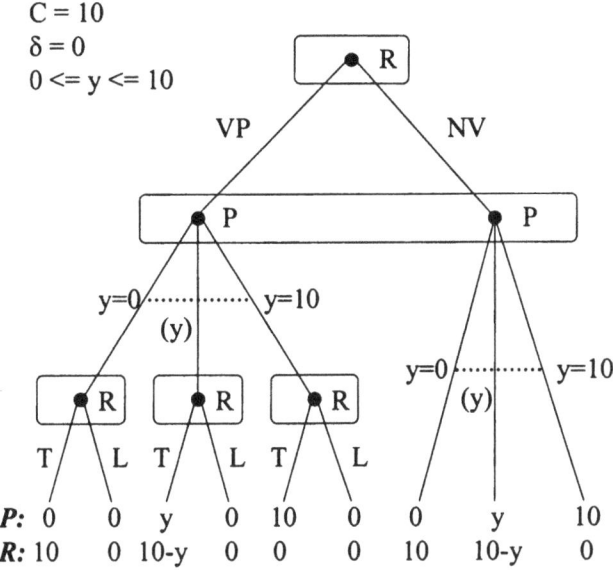

Figure 11: Game-Tree for Design C with a Bonus of Zero

4. The Design with a Low Constant Bonus: D

In contrast to designs A and B, the bonus for design D is held constant at DM 0,50. This low bonus size was chosen since a fair division of the cake and a proportional bonus of 10% would produce a bonus amount of exactly DM 0,50. But with a lower share for the receiver, the bonus amount decreases in designs A and B, and becomes zero in case of a zero share. In designs with a constant bonus, the receiver obtains the full bonus amount in case of no veto power, even if his share is as low as zero. This is illustrated by the game tree for design D in Figure 12.

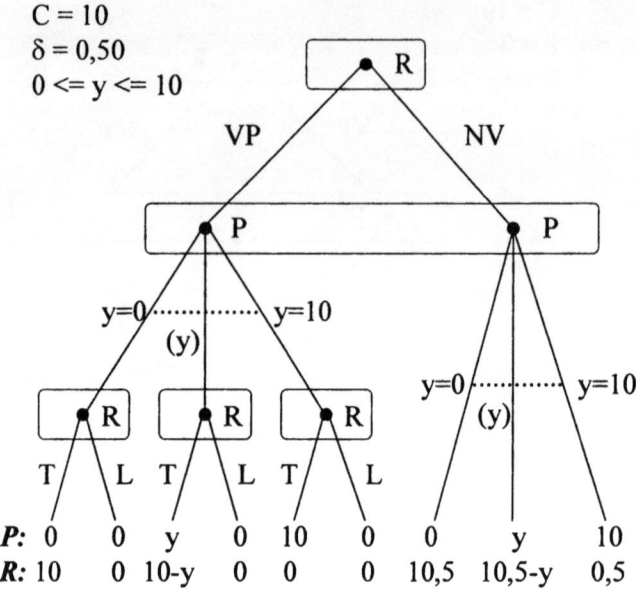

Figure 12: Game-Tree for Design D with a Bonus δ of DM 0,50

5. The Design with a High Constant Bonus: E

The game tree for design E with a high constant bonus of DM 2,00 is nearly identical to the design D game tree, with the size of the bonus as the only difference. The bonus size of DM 2,00 was chosen to generate a distinctively higher incentive than the low bonus, but it is also kept far below the DM 5,00 that represent an equal split.

6. Designs with a Constant Price: F, G and H

The game tree for design F is identical to the one for design D. The difference between these designs is produced by the wording of the instructions. Instead of paying a bonus amount to receivers who abandon veto power, all receivers obtain this bonus amount from the start as an endowment. To receive veto power, a price of the same size has to be paid. As expected, the only difference between designs F, G, and H is the size of the bonus. In the following chapter, some options for alternative designs are discussed, and after that the experimental results for the FTP game are presented in chapter G., including hypotheses about behavior and differences between designs.

II. Alternative Designs

The basic FTP game model can be implemented using various different designs. The approach for this study has already been outlined, but there are of course many alternatives. Some additional ideas are shortly introduced in this chapter and should only serve as a reference or inducement for future research. The strategy method was already discussed in chapter B.II.1. above, including possible discrepancies in behavior between spontaneous play and strategy play. The FTP game is not played with the strategy method, but the experiments for the RAP game will be implemented by using the strategy method.

A lot of additional designs could be produced by changing the bonus sizes. But the three developed bonus values of zero, low and high cover a wide range. It was already mentioned that design G with a bonus between a high and a low bonus at DM 1,- did not have to be realized, but it could of course be run to receive a broader data set. A high proportional bonus is also possible, for example 40%, which would amount to DM 2,- in case of an equal split of the DM 10,- cake, and therewith correspond to the high constant bonus of DM 2,-. A third and new alternative is an extremely high bonus of 100% or DM 5,- (or even DM 10,-). This might be interesting to test whether some receivers would still stick to their veto power in this case, indicating a high motivation to punish unfair proposers.

An auction mechanism also offers certain possibilities. The veto power could be auctioned to those receivers whose bids are in the top half of all receivers' bids. This should produce an indifference value for veto power, i.e. the lowest bid that still bought veto power. Alternatively, participants could just be asked what they would like to pay for veto power. After that, a price for veto power could be determined by chance. This should also result in a representative price for veto power. Furthermore, the proposer role could be auctioned to those participants whose bids are in the top half of all bids, leaving the receiver role to the rest. But this would probably have a strong effect on demand behavior, and therewith also on veto power decisions. A final thought should be given to detailed questionnaires. By asking the right questions, insights into the underlying motivational forces, for example intrinsic motivation or fairness motives, could be gained. But such questions would be open questions, and the answers would be hard to analyze or compare. Nevertheless, all of these alternative procedures should be considered for future research. In the next chapters, the decision data is browsed and analyzed.

G. Experimental Results for the FTP Game

In this chapter, the experimental results for the FTP game are analyzed. In the first part I., the outcomes of the different decisions are simply listed in an appropriate format to give an overview and to gain a first impression. The following chapters II. and III. contain the statistical analysis, and especially some hypotheses testing. Paragraph IV. summarizes the results of the FTP game.

I. An Overview of the Decisions in the FTP Game

Referring to the three stages of the FTP game, the basic experimental data is presented in three parts. The veto power decisions are shown first, followed by the demands and the acceptance decisions. While tables distinguish between designs, the charts are based on aggregated data only, since a more detailed analysis along with some statistical tests follow in chapter G.III. The expectations of the other player about the respective decision of a certain participant are also listed. All collected expectations were based on questions. The answers to these questions were neither controlled nor accompanied by incentives. But since the questions had to be answered at the same time and on the same sheet as the final decisions, the answers should have a certain quality. Nevertheless, the expectations are only used to provide some additional insights and to support a reliable behavioral theory. The main focus remains on the real decisions. For design A, no expectations were recorded. In the other designs, some participants (2%) did not answer the questions about their expectations.

1. The Veto Power Decisions

Figure 13 shows the frequency of veto power and expected veto power decisions for all designs. The difference between design C and the rest of the designs is obvious. Without design C, 50% of the proposers expected a veto power decision. Furthermore, the discrepancy between the expectations of the proposers and the final decisions of the receivers appears to be interesting.

Design	Veto Power	No Veto Power	Veto Power Expected	No Veto Power Expected
Design A	7	13	-	-
Design B	4	16	7	13
Design C	17	3	18	2
Design D	4	16	11	9
Design E	2	18	6	14
Design F	9	11	13	7
Design H	5	15	13	7
Sum	48	92	68	52

Figure 13: Realized and Expected VP Decisions for all FTP Designs

Without exception, all designs show a higher number of proposers expecting veto power decisions than the number of receivers finally choosing veto power. This is also illustrated by the two pie charts below.

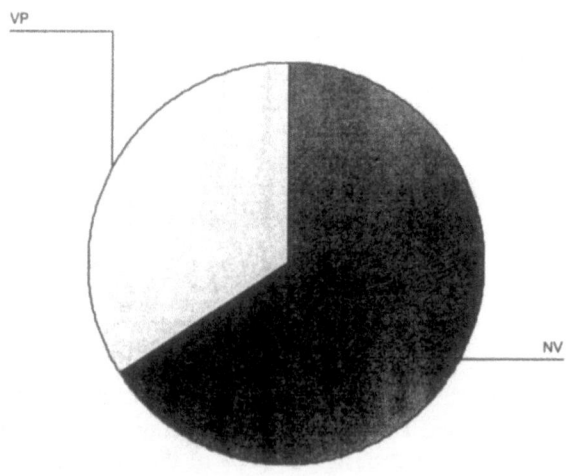

Figure 14: Veto Power Decisions for all FTP Designs

I. An Overview of the Decisions in the FTP Game

While Figure 14 shows that the majority of receivers (66%) decided to abandon their veto power, this was not anticipated by proposers, who expected only 43% to refrain from veto power, as shown in Figure 15 below. The missing values are caused by design A only.

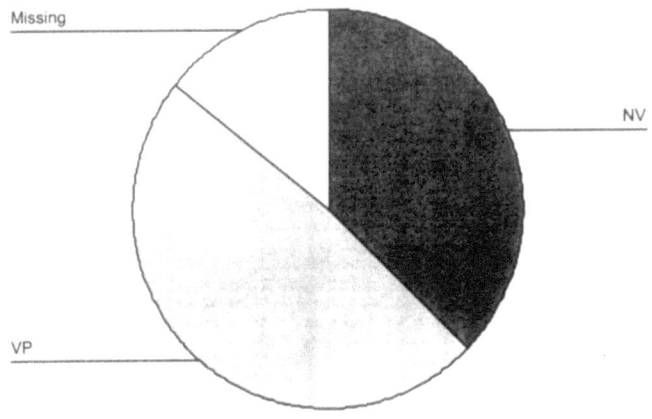

Figure 15: Expected Veto Power Decisions for all FTP Designs

2. The Proposals

The following demands of the DM 10,00 cake were made, see Figure 16. As expected, the equal split at DM 5,- was the most frequent choice with 30%. In line with considerations of prominence, demands of other full DM amounts at 6, 7, 8 and 10 DM were also made often, as well as DM 5,25 and DM 5,50. The game theoretic solution seems to play a certain role, since demands of the whole cake or nearly the whole cake add up to 10%, including offers of DM 0,01 and 0,10. An offer of DM 0,10 can be interpreted as being the lowest possible offer due to considerations of perception.

For design A, 21 proposals were recorded, but one player could not be matched with a receiver since only 41 players were present. This means that the acceptance decision of step 3 does not exist in this case. The respective demand

Demand	Des. A	Des. B	Des. C	Des. D	Des. E	Des. F	Des. H	Sum
4,50			1					1
4,90	1							1
5,00	8	3	9	8	5	5	5	43
5,24		1						1
5,25				3		2		5
5,42	1							1
5,46		1						1
5,50	3			1		4		8
5,55	1							1
5,75	1							1
5,90		1						1
5,99			1					1
6,00	3	3	6	4	10	2	7	35
6,20		1						1
6,50				2				2
6,66							1	1
7,00		2		1	2	1	2	8
7,50		1	1			1	1	4
8,00	1	2		1		1	1	6
8,90						1		1
9,00		1				1		2
9,50		1				1		2
9,90		1						1
9,99		1	1				1	3
10,00	2	1	1		3	1	2	10
Sum	21	20	20	20	20	20	20	141
Average	5,91	7,08	5,95	5,66	6,45	6,47	6,66	6,31
Median	5,50	6,60	5,495	5,25	6,00	5,50	6,00	6,00
Modus	5,00	5 and 6	5,00	5,00	6,00	5,00	6,00	5,00
	1,52	1,79	1,55	0,83	1,64	1,71	1,67	1,59

Figure 16: Demands for all FTP Designs

I. An Overview of the Decisions in the FTP Game   89

was DM 5,-. The total average demand of DM 6,31 is close to the two third demand observed in many previous experiments as described in chapter B.III. above.

The modus of most designs is DM 5,-, with DM 6,- being the modus for two designs and design B showing both values 5,- and 6,- as modus. This tendency towards a bimodal distribution is illustrated by Figure 17, which includes all designs.

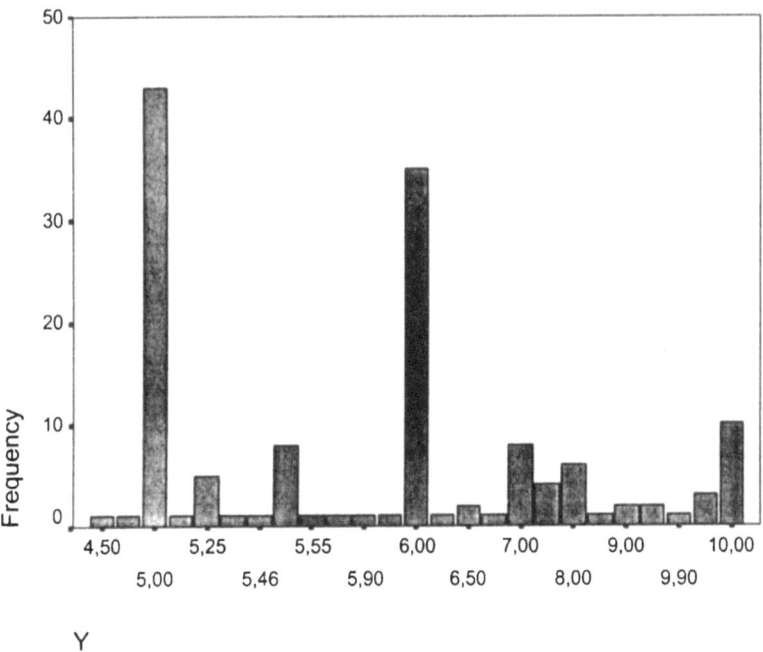

Figure 17: Distribution of Demands for all FTP Designs

The expectations of the receivers about the demands follow a similar pattern and are listed in Figure 18 below. A general difference is obvious, since the expected values are consistently lower. With an average expected demand of DM 5,71 versus a DM 6,31 average real demand, the difference amounts to 9%. But the equal split is again the dominant value (44%), with DM 4, 6, 10 and 5,50 also expected frequently.

G. Experimental Results for the FTP Game

Expected Demand	Design B	Design C	Design D	Design E	Design F	Design H	Sum
3,50			1		1		2
4,00	1	1		4		4	10
4,50		2		1			3
4,75					1		1
4,99			1				1
5,00	10	12	8	8	7	7	52
5,25			2				2
5,50	1	1	3		6		11
5,55	1						1
6,00		3		4	1	2	10
6,15			1				1
6,50			1		2		3
7,00				3	1	1	5
7,25						1	1
7,50	1	1					2
8,00						1	1
8,50	1		1				2
9,00						1	1
9,90	1						1
10,00	3		2			3	8
Sum	19	20	20	20	19	20	118
Average	6,37	5,20	5,83	5,28	5,38	6,21	5,71
Median	5,00	5,00	5,125	5,00	5,50	5,00	5,00
Modus	5,00	5,00	5,00	5,00	5,00	5,00	5,00
σ	2,16	0,73	1,70	0,99	0,77	2,12	1,57

Figure 18: Expected Demands for all FTP Designs

I. An Overview of the Decisions in the FTP Game

The expectations about the demands include less different values than the real demands. Therefore, the distribution of expected demands in the following Figure 19 is less diversified.

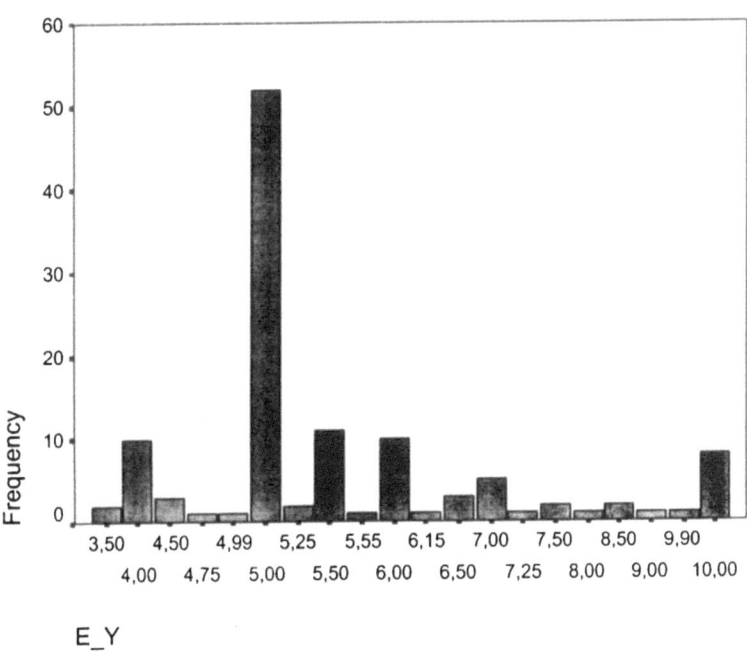

Figure 19: Distribution of Expected Demands for all FTP Designs

For designs B to H, the expectations of the receivers about the demand of the proposers were collected. The averages of the real demands and the expected demands for all of these designs are included in Figure 20. Comparing the expectations with the real average demands leads to the result that the average expectations are too optimistic in all but one case, which is design D. The average demand of design D with DM 5,66 is the lowest of all designs. For all other designs, the receivers expected a lower demand on average than finally occurred.

Design	Average Demand	σ	Av. Expected Demand	σ	Difference
B	7,08	1,79	6,37	2,16	0,71
C	5,95	1,55	5,20	0,73	0,75
D	5,66	0,83	5,83	1,70	-0,17
E	6,45	1,64	5,28	0,99	1,17
F	6,47	1,71	5,38	0,77	1,09
H	6,66	1,67	6,21	2,12	0,45
Overall	6,38	1,60	5,70	1,57	0,68

Figure 20: Average Demanded and Expected Shares

Another approach can be taken by analyzing the differences between receivers who have chosen veto power and those who have not. The difference between the expected demands of these two groups of receivers is shown by Figure 21.

Design	Average Expected Demand VP	σ	Average Expected Demand NV	σ
B	8,33	2,89	6,00	1,88
C	5,26	0,73	4,83	0,76
D	5,37	0,75	5,95	1,87
E	6,00	1,41	5,19	0,96
F	5,56	0,62	5,25	0,86
H	7,60	2,51	5,75	1,84
Overall	5,91	1,61	5,61	1,55

Figure 21: Average Expected Demands for VP and NV Choices

It should be mentioned that the calculated averages of Figure 21 are often based on just a few values, leading to wrong impressions about their real impact. The overall averages appear to be more reliable figures. Comparing the overall average expected demands for the VP and the NV choices, the expected demand of receivers who have chosen veto power (DM 5,91) is slightly higher than the corresponding value for receivers who have refrained from veto power (DM 5,61).

A last look should be given to the average expected demand for the NV choices in the zero bonus design C. NV choices should have happened less frequently in design C than in other designs with a positive bonus. Even though this seems to be the case, some decision makers have chosen NV despite the fact that there was no bonus available for refraining from veto power in design C. The unrealistic average expectation about the demand of DM 4,83 might be one of the causes of the three NV choices in design C, but this question is impossible to answer here. However, a receiver can hardly expect to obtain more than half of the cake.

3. The Acceptance Decisions

The number of recorded acceptance decisions is reduced by the fact that the majority of receivers refrained from veto power, therewith also excluding a step 3 decision. Only 48 of the possible 140 acceptance decisions finally had to be made, as shown in Figure 22.

Design	Accepted	Rejected	Expected Acceptance	Expected Rejection
Design A	7	0	-	-
Design B	3	1	12	7
Design C	13	4	19	1
Design D	3	1	15	3
Design E	1	1	15	2
Design F	7	2	15	5
Design H	5	0	13	7
Sum	39	9	89	25

Figure 22: Table of Acceptance Decisions for all FTP Designs

G. Experimental Results for the FTP Game

A total of 9 or 19% rejections took place, while 22% were expected, see the following two pie charts, Figure 23 and Figure 24.

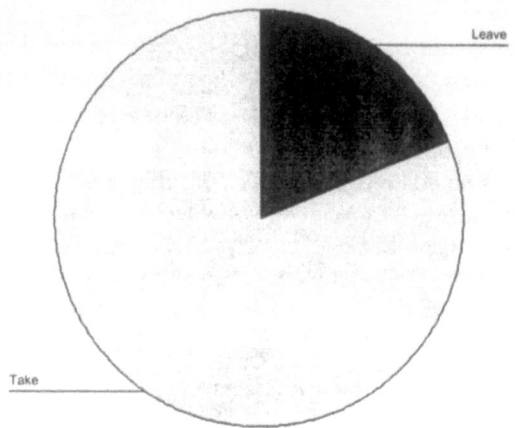

Figure 23: Acceptance Decisions for all FTP Designs

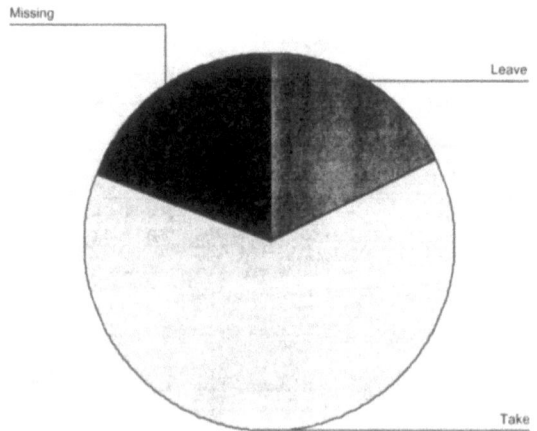

Figure 24: Expected Acceptance Decisions for all FTP Designs

I. An Overview of the Decisions in the FTP Game   95

The expectations about rejections do not differ extremely from the real decisions. While 25 of 114 proposers or 22% expected their demand to be rejected in case of the existence of veto power (see Figure 24), 9 of 48 demands or 19% were finally rejected in stage 3 of the FTP game (see Figure 23). The remaining question is why 25 proposers made demands that they expected to be rejected. The pure existence of expected rejections appears to be astonishing, since a proposer who expects his demand to be rejected should simply demand less. A simple explanation is offered by the fact that veto power could be abandoned by receivers and frequently was. Following that, an expected rejection would exist together with a corresponding expectation of no veto power by the respective proposer. But of the 25 proposers who expected a rejection of their demand, 13 expected a veto power choice. Accidentally, none of these 13 demands was rejected. These and other questions are explored in the statistical part in chapter G.III.

4. Payoffs and Efficiency

Depending on their decisions, the participants reached certain payoffs, and even though the cake had the same size for all designs, the varying bonuses lead to different average payoffs as well as to different levels of efficiency. Figure 25 presents the average payoffs for proposers and receivers according to each design. The average payoff of the proposer was lowest in design C, where no bonus existed, and also far lower than the overall average payoff. This is caused by the fact that a lot of receivers have chosen veto power in design C, and several of them finally rejected the demand, leading to zero payoffs in four cases. On the other hand, the average payoff of design C for the receiver is not the lowest of all designs, and just a little lower than the overall average payoff.

The efficiency for each design is calculated using the sum of the average payoffs for proposer and receiver. This sum is divided by the total possible payoff, which includes a bonus. The bonus is always considered, no matter whether veto power was chosen or not. For designs A and B, the maximum possible bonus of DM 1,- in case of a 100% share for the receiver was used. Again, design C proves to be least efficient with a level of 80%. Design H produced the highest efficiency with a level of 95%, even though the high bonus of DM 2,- was not reached in 5 of 20 cases. Just slightly lower levels of efficiency were reached in designs A, D, and E, while designs B and F finish with 88%, just a little worse than the overall efficiency level of 90%. Therewith, a bonus seems to promote efficiency, since only design C without a bonus stayed clearly below the average efficiency level. The main reason for this is again the higher number of rejections. The following statistical analysis provides further insights into these problems and shows whether these first impressions can be proven by means of statistical testing.

Design	Proposer	σ	Receiver	σ	Efficiency
A	5,96	1,55	4,28	1,64	93,09 %
B	6,78	2,39	2,93	2,05	88,27 %
C	4,27	2,24	3,73	1,97	80,0 %
D	5,36	1,51	4,54	1,35	94,29 %
E	6,15	2,18	5,15	2,11	94,17 %
F	5,71	2,45	3,57	1,87	88,38 %
H	6,66	1,67	4,84	1,91	95,83 %
Total	5,84	2,14	4,14	1,96	90,58 %

Figure 25: Average Payoff Overview (in DM)

II. Design Background and Hypothesis Approach

The design structure for the FTP game is aimed at analyzing the impact of the bonus on behavior. But the general tendency of the veto power decisions also provides chances to gain further insights into bargaining behavior. Conclusions about the impact of the receiver's possibility to influence the rules of the bargaining process might be drawn. To support this, the general approach of hypothesis testing is used, which is illustrated in this chapter. The hypotheses are formulated, tested and interpreted in the following chapter G.III.

The bonus most likely influences the decisions. For example, it has to be explored under which circumstances the receivers sell their secret veto power, and whether significant differences between certain bonus types exist. This might lead to some more insights about the importance of freedom of choice and could even serve as a proof for its relevance. The first step is to compare the situation without a bonus with those situations in which a bonus was available. It could be assumed that the participants are more willing to exclude alternatives if they are paid for doing so.

To find a proof for this theory, a hypothesis test can be performed. According to Greene (1993), a formal test procedure requires the formulation of a null hypothesis and a hypothesis with the exact opposite statement, the so-called alternative hypothesis. The null hypothesis can then be tested by means of a statistical procedure, which dictates whether the null hypothesis is rejected or not. The rejection of the null hypothesis then serves as a statistical proof for the alternative hypothesis (given a certain error probability p, which is also called the significance level). In the following chapter III., all null hypotheses referring to the FTP game are labeled H^0_{FTPX}, where X is the number of the hypothesis. Alternative hypotheses are labeled H^A_{FTPX}, respectively. By performing certain statistical tests of these hypotheses, the impact of freedom of choice can be evaluated. This is done in the next chapters.

III. Statistical Analysis for the FTP Game

In this chapter, selected characteristics of the collected data are presented to illustrate the major effects that have been observed. Several hypotheses are developed, applied and tested. While hypotheses H^0_{FTP1}, H^0_{FTP2}, H^0_{FTP3}, H^0_{FTP4}, and H^0_{FTP5} deal with veto power decisions, hypotheses, H^0_{FTP6}, H^0_{FTP7}, H^0_{FTP8}, and H^0_{FTP9} are focusing on the proposals. H^0_{FTP10} takes the acceptance decisions into account. These and other major impacts are shown and discussed in the paragraphs below, starting with the veto power decisions.

1. The Veto Power Decisions

Considering freedom of choice, the veto power decisions of the receivers have the major impact on their own choice situation. Therefore, any effects in the frequency of veto power choices caused by the different designs should be analyzed in detail. Starting with general remarks and a short overview, several hypotheses are developed and tested. A graphical illustration finalizes the investigation of the veto power choices.

G. Experimental Results for the FTP Game

a) General Tendencies for the Veto Power Decisions

The frequency of the veto power choices in percentages for all designs is shown in the following Figure 26. Overall, no veto power was chosen about twice as much as veto power, which is a rather surprising result according to considerations of fairness and punishment. After excluding design C, the frequency of NV choices reaches 76%. The shaded design C is especially interesting, since it was played with a bonus of zero, while all other designs had a positive bonus for choosing NV. Design C is the only setting in which more VP than NV occurred. Therefore, a first testing focus is aimed at design C.

Design	Observations	Veto Power	No Veto Power	Veto Power in %	No Veto Power in %
Design A	20	7	13	35	65
Design B	20	4	16	20	80
Design C	20	17	3	85	15
Design D	20	4	16	20	80
Design E	20	2	18	10	90
Design F	20	9	11	45	55
Design H	20	5	15	25	75
Overall	140	48	92	34	66
Overall w/o Design C	120	31	89	24	76

Figure 26: Veto Power Choices for all FTP Designs

b) Analysis of the Veto Power Decisions

The hypothesis testing procedure explained above will be applied for the veto power choices in the different designs. In the context of a bonus, a possible first null hypothesis for the FTP game could be formulated as follows.

III. Statistical Analysis for the FTP Game 99

H^0_{FTP1} A bonus of any kind does not lead to more NV choices than a zero bonus.

The corresponding alternative hypothesis is H^A_{FTP1}:

H^A_{FTP1} A bonus of any kind leads to more NV choices than a zero bonus.

To test H^0_{FTP1}, the Pearson Chi square test of independence is used, see Siegel (1956). This test determines whether two variables are independent. In this case, the independence of the variables "size of the bonus" and "no veto power choice" is tested. To perform this task, the expected number of cases in each cell is determined, and then compared to the observed number of cases. If the difference between these figures is significant, the null hypothesis can be rejected. One of the conditions for this test is that the expected number of cases must not be too small. All expected frequencies should be at least 5. Authors like Everitt (1977) state that this strict condition can usually be relaxed, but this line of investigation is of no further interest for this study. In case of expected frequencies smaller than 5, Fisher's exact test can and will be used instead.

Another thought should be given to the question whether a one-tailed or a two-tailed test should be used. Generally, a two-tailed test is used if the null hypothesis states that a certain parameter has a particular value. In contrast to that, a one-tailed test may be used if the null hypothesis states that a parameter is not above (or below) a certain value. Therefore, hypothesis H^0_{FTP1} is a one-tailed hypothesis, and a one-tailed test approach can be used. Most of the hypotheses in this study are formulated in a way that allows for one-tailed testing. The results for the following Chi square tests were approximated based on a two-tailed significance level. One-tailed significance levels are given whenever possible. Fisher's exact test was performed one-tailed if appropriate.

The following result was obtained regarding H^0_{FTP1}. According to Figure 26, only 3 of 17 players have chosen NV in design C, but in the presence of any kind of bonus, 89 of 120 participants played NV. This difference is highly significant ($p < 0.001$, $\chi^2 = 26.636$) and leads to the rejection of H^0_{FTP1}. The Chi square test can be used, since the minimum expected frequency of 6.86 is greater than 5. As could have been expected, Fisher's exact test is also significant ($p < 0.001$, one tailed). Figure 27 below shows the results of the Chi square test and Fisher's exact test (one-tailed) for H^0_{FTP1} based on the results of design C vs. the results of all other designs. Regarding the applicability of the

G. Experimental Results for the FTP Game

Chi square test, all expected frequencies are greater than 5. The differences between design C and each of the other designs are also highly significant.

Name of test: Design C versus:	Chi square test		Fisher's exact test
	p	χ^2	p (one-tailed)
Design A	0.001	10.41	0.002
Design B	< 0.001	16.94	< 0.001
Design D	< 0.001	16.94	< 0.001
Design E	< 0.001	22.55	< 0.001
Design F	0.008	7.03	0.009
Design H	< 0.001	14.54	< 0.001

Figure 27: Test Results for Hypothesis H^0_{FTP1}

Based on these test results, H^0_{FTP1} can be rejected for all possible comparisons, therewith supporting H^A_{FTP1}. A bonus of any kind leads to more NV choices than a zero bonus. In other words, the difference in the frequency between design C and the other designs clearly indicates the importance of the alternative to punish (rejecting the offer instead of accepting it). Without a bonus, only a few participants were willing to give up their veto power. This changes dramatically when monetary incentives are applied – the majority of players refrained from their veto power.

More specifically, it can be tested whether veto power is the dominating choice in case of a zero bonus (design C). To test this, the following null hypothesis H^0_{FTP2} is formulated.

H^0_{FTP2} In the zero bonus situation of design C, decision makers do not chose veto power (VP) more frequently than no veto power (NV).

The corresponding alternative hypothesis is H^A_{FTP2}:

H^A_{FTP2} In the zero bonus situation of design C, decision makers chose VP more frequently than NV.

To test H^0_{FTP2}, a one-sample Chi square test can be performed. This approach, which is different from the Chi square test of independence used in connection with H^0_{FTP1}, tests whether the observed frequencies are different from the expected frequencies. In this case, H^0_{FTP2} implies that veto power and no veto power were chosen equally frequent, which would be 10 times out of the 20 existing cases for design C. Again, that test can be used in this situation since the number of expected frequencies (10) is clearly above 5. The observed frequencies are different from the expected frequencies, since 3 of 20 receivers abandoned their veto power even though no bonus was offered. This difference is highly significant ($p = 0.002$, $\chi^2 = 9.8$), leading to the rejection of H^0_{FTP2}. This means that receivers attach a certain value to the right to punish. Even though the choice for veto power is not signaled towards the proposers, most receivers have chosen to have veto power. However, some receivers nevertheless abandoned their veto power.

The existence of no veto power choices when no bonus is achievable might indicate that some receivers have no intention to punish at all. Considering the arguments about fairness and annoyance discussed in chapters C.III. and C.V. above, these receivers might be motivated by a fear of annoyance. They exclude the punishment situation, since they anticipate that a rejection decision is not only caused by annoyance, but also causes more annoyance itself. A negative outcome as such might be the more frustrating the longer someone has to think about it. Therefore, they avoid the act of rejecting by excluding veto power and therewith implicitly accepting all offers.

Building up on this, it has to be analyzed whether a high bonus generates a stronger motivation to refrain from veto power than a low bonus. In general, receivers could simply try to secure the bonus for themselves, and a higher bonus might reinforce this behavior. The corresponding null hypothesis is H^0_{FTP3}:

H^0_{FTP3} A high bonus does not lead to more NV-choices than a low bonus.

The resulting alternative hypothesis reads as follows:

H^A_{FTP3} A high bonus leads to more NV-choices than a low bonus.

Therefore, the low bonus design D has to be compared with the high bonus design E. In combination with that, the low price design F and the high price design H can also be compared. In both cases, NV was chosen more often in the high (bonus or price) condition than in the low one. But since the difference is only two (design D vs. E), respectively four more NV decisions (design F vs. H), this effect does not prove to be significant on a high level. For designs D s. E, Fisher's exact test delivers a significance level of $p=0.331$, and for designs F vs. H p is 0.160. For these samples alone, H^0_{FTP3} can not be rejected.

Combining the bonus and price payment designs raises the number of observations for a test procedure, but includes some risk of falsification due to uncontrolled influences. Nevertheless, it appears to be plausible to compare all decisions made in the presence of high monetary incentives with those made at low incentives. Confronted with high incentives (designs H and E), 33 out of 40 participants have chosen NV, while only 27 out of 40 have done so in the presence of low incentives (designs D and F). Using Fisher's exact test, this difference is weakly significant ($p = 0.098$, one tailed). On a significance level of 10%, H^0_{FTP3} can be rejected for the combined data of the price and bonus payment designs.

The underlying problem regarding the significance level is of course the small sample, but also the fact that with already a high percentage of NV choices in low bonus settings, there is not much room left for a further increase, for example the percentage of NV choices amounts to 80 in design D with a low bonus, and increases to 90% in design D with a high bonus. However, there is some evidence to support H^A_{FTP3}.

Some thoughts should be spent on the difference between the bonus payment and the price payment designs. It could be assumed that a monetary endowment like the additional show-up fee of the price payment designs has the effect that a lot of participants want to keep this amount instead of spending it for veto power. For the endowment effect and additional literature, see for example Tietz (1992, 1999). Following that, less VP choices could be expected in the price payment designs.

Contrary to that, the low as well as the high bonus payment design produced less VP choices than the corresponding price payment designs (20% and 10% vs. 45% and 25%). To confirm the significance of this effect, the following null hypothesis is formulated:

H^0_{FTP4} In the price payment designs F and H, VP choices do not happen more frequently than in the bonus payment designs D and E.

III. Statistical Analysis for the FTP Game 103

The respective alternative hypothesis is H^A_{FTP4}.

H^A_{FTP4} In the price payment designs F and H, VP choices happen more frequently than in the bonus payment designs D and E.

The null hypothesis H^0_{FTP4} is tested by means of a Chi square test of independence. This test is significant for the pooled data of designs D and E versus F and H ($p = 0.039$, $\chi^2 = 4.26$), and also weakly significant for design D vs. F ($p = 0.091$, $\chi^2 = 2.84$), leading to the rejection of hypothesis H^0_{FTP4}. Generally, veto power appeared to be more attractive in case receivers had to pay for it. Whether this observation is contrary to the common endowment effect, as demonstrated by Tietz (1992, 1999), has to be clarified by future research.

Turning to the expectations about veto power choices, it is remarkable that proposers in all designs expected more veto power choices than finally happened. A corresponding null hypothesis is H^0_{FTP5}:

H^0_{FTP5} Proposers do not expect more veto power choices than finally happen.

The respective alternative hypothesis is H^A_{FTP5}.

H^A_{FTP5} Proposers expect more veto power choices than finally happen.

The null hypothesis H^0_{FTP5} is tested by means of a Chi square test of independence. H^0_{FTP5} can be rejected based on the data of design D ($p = 0.022$, $\chi^2 = 5.22$) and design H ($p = 0.011$, $\chi^2 = 6.46$). Over all designs, 48 receivers have chosen veto power, and 92 refrained from it by choosing NV. The proposers expected veto power 68 times, and only 52 NV decisions. This difference is also significant on a high level ($p < 0.001$, $\chi^2 = 13.09$), and H^0_{FTP5} can also be rejected based on the pooled data of all designs. Therewith, proposers appear to give a higher attention or value to a veto power decision than receivers do. This might be caused by different expectations about the demands, which are analyzed in more detail in paragraph G.III.2.b. If receivers expect lower demands than proposers are finally willing to make, this could explain the difference between veto power expectations and realized veto power decisions.

c) Graphical Illustration of the Veto Power Decisions

Several effects on the frequency of veto power choices have been shown in the last paragraph, caused by the size of the bonus and the method by which this bonus was implemented. There were four different bonus variations:

- No bonus.
- Proportional bonus: a bonus of 10 % was added to the personal payoff amount of the responder.
- Constant bonus: a bonus of DM 0,50 (or 2,-) was paid by the experimenter to the responder after abandoning his veto power.
- Constant price: a price of DM 0,50 (or 2,-) was paid by the responder to the experimenter after choosing veto power. The responder received a bonus amount of the same sum at the start of the experiment. Therewith, by refraining from veto power, the responder could keep that bonus, otherwise he had to give it back.

The following Figure 28 illustrates all designs with the percentages of no veto power choices (NV) and the corresponding bonus types. The zero bonus design C produces all three symbols at a bonus size of zero, since it can be compared to all proportional bonus, constant bonus and constant price designs. Designs A and B had a proportional bonus of 10%. The monetary value of that bonus is measured by using the average expected share of the receivers of DM 3,63. This leads to an average expected bonus of DM 0,36 for design B. Since expectations were not inquired in design A, only design B is included in Figure 28. But the realized average bonus amounts of DM 0,41 for design A were pretty close to the expected value of DM 0,36 for design B. Figure 28 visualizes the main effects of the various bonuses which have already been tested and proved to be significant with regards to hypotheses H^A_{FTP1} to H^A_{FTP4}.

Even a marginal bonus amount of DM 0,36 or DM 0,50 leads to a strong increase of NV choices (H^A_{FTP1}). This effect is reinforced by a higher bonus (H^A_{FTP3}). Apart from that, the bonus payment mechanism has a stronger impact than the price payment mechanism (H^A_{FTP4}). Following H^A_{FTP5}, proposers appear to give a higher attention or value to a veto power decision than receivers do. This might be caused by different expectations about the demands. The demands as well as the expected demands are analyzed in the next chapter.

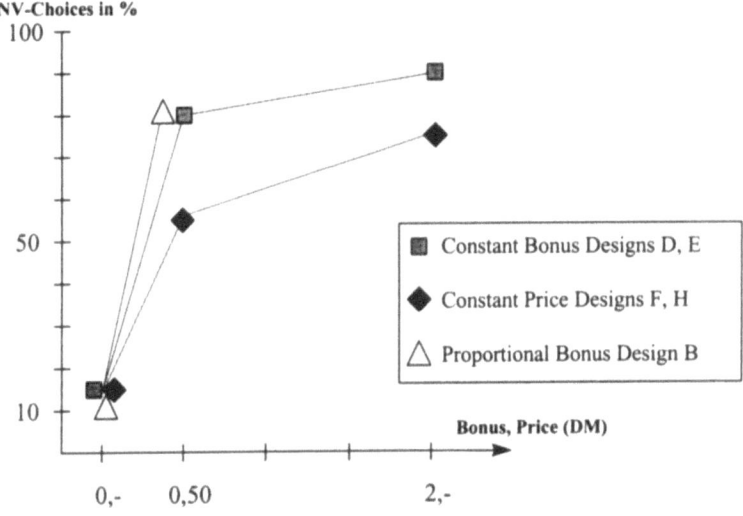

Figure 28: Percentages of NV Choices for all Bonus Types

2. The Proposals

Considering the demands of the proposers, it appears to be interesting whether these demands follow the regularities observed for the veto power choices, or whether there are some other unique effects. First, a general impression is given, followed by hypotheses testing and a graphical illustration.

a) General Tendencies for the Demand Decisions

The distribution of the proposals in each of the seven designs was already demonstrated in chapter G.I.2. above. For this paragraph, it appears to be appropriate to exclude design A from all tests regarding the demands, since the monetary incentives in this design differed from the other designs. Only for design A, a random payoff procedure with a probability of 0.25 for being paid was used. Figure 29 shows the average demands for all relevant designs.

Design Type	Average Demand [DM]	σ
No bonus (Design C)	5,95	1,55
Bonus 0,50 (Design D)	5,66	0,83
Bonus 2,00 (Design E)	6,45	1,64
Price 0,50 (Design F)	6,47	1,71
Price 2,00 (Design H)	6,66	1,67
Bonus 10% (Design B)	7,08	1,79
Overall Average	6,38	1,60
Average w/o Design C	6,46	1,61

Figure 29: Average Demands for Designs B to H

b) Analysis of the Demand Decisions

The impact of the bonus on the behavior of the receivers was already discussed, see paragraph G.III.1.b. Another question is whether such behavior of the receivers is anticipated by proposers. Generally, a bonus led to more NV choices. If proposers anticipate this kind of behavior, they might expect more NV choices in bonus situations than in the zero bonus setting. Therefore, they might demand more money in the bonus settings compared to the no bonus situation. The reason behind this could be the fact that proposers would have a lower fear of rejection due to the smaller expected number of veto power that has to be faced in bonus settings. This leads to the null hypothesis H^0_{FTP6}.

H^0_{FTP6} Demands in the bonus settings are not higher than demands in the zero bonus setting.

The corresponding alternative hypothesis is

H^A_{FTP6} Since a bonus of any kind leads to more NV decisions than a zero bonus, and this is anticipated by proposers, the demands are higher in the bonus settings than in the zero bonus setting.

III. Statistical Analysis for the FTP Game

For all but one of the bonus designs, the average demands are higher than the DM 5.95 of the no bonus design C (see Figure 29). This single exception is design D with a constant bonus of DM 0.50, which produced an average demand of DM 5.66. The distribution of demands for design D, as already shown in chapter G.I.2., provides the reason for this divergence. Only in design D, no demands of more than DM 8.00 took place, while all other designs include at least one demand of the maximum DM 10.00. Additionally, a value of DM 5.66 is not dramatically out of scope, and still clearly above an equal split. To test H^0_{FTP6}, a one-tailed Mann-Whitney-U-Test can be performed. This test ranks the demands of two samples, and calculates the average rank of each sample. Then it analyzes whether the difference between these two average ranks is significant. The difference between the demands of designs B (average rank of 24.75) and C (average rank of 16.25) proves to be significant (Mann-Whitney-U-Test, $p = 0.009$, one-tailed, $Z = -2.34$), leading to a rejection of H^0_{FTP6}.

Furthermore, this also applies for a comparison between the zero bonus design C and all other designs B to H. The 20 demands of design C average at DM 5.95, while the other 100 demands, all made in the presence of a bonus, turn out to have a higher average of DM 6.46. The difference between the two groups of cases is significant (MWU-Test, $p = 0.028$, one tailed, $Z = -1.90$). On the basis of the comparison between designs B and C as well as between all designs and design C, H^0_{FTP6} can be rejected, therewith giving support to H^A_{FTP6}. Proposers do in fact demand more money in bonus settings than in the zero bonus setting. The next step is to compare demands in high and low bonus situations.

A similar assumption can be made for the difference between high and low bonus situations. Proposers might demand more money in high bonus situations, due the lower expected number of VP choices and the resulting lower fear of rejection, see null hypothesis H^0_{FTP7}:

H^0_{FTP7} Demands in the high bonus settings are not higher than demands in the low bonus settings.

The respective alternative hypothesis is:

H^A_{FTP7} Since a high bonus leads to more NV decisions than a low bonus, and this is anticipated by proposers, the demands are higher in high bonus situations.

Of course, proposers might also just want to have a part of the high bonus of DM 2,00. This could be interpreted towards a crowding-out of fairness by monetary incentives on behalf of proposers. But this might also be anticipated by receivers, who then keep their veto power to be able to punish extreme demands. In this case, trust in fairness on behalf of the receivers would be crowded out by that higher bonus.

The tendency implied by hypothesis H^A_{FTP7} is plain to see by comparing the mean demands included in Figure 29. While design D with a low bonus of DM 0.50 shows an average demand of DM 5.66, the high bonus design E reaches an average demand of DM 6.45. Design F with a low price of DM 0.50 has an average demand of DM 6.47, the high price design H produces an average demand of DM 6.66. Regarding H^0_{FTP7}, a MWU test is performed for the pooled data of the low bonus or price designs D and F (average rank of 36.22) versus the high bonus or price designs E and H (average rank of 44.78). This test turns out to be weakly significant (MWU-Test, $p = 0.046$, one tailed, $Z = -1.68$). Based on this test result, H^0_{FTP7} can be rejected, indicating that the demands are higher in high bonus and price situations. Another observation is that the constant price designs F and H seem to produce higher average demands than the constant bonus designs D and E. However, this difference did not prove to be significant.

A proportional bonus of 10% seems to sound more attractive than all constant bonuses and prices, since the average demand is higher than in all four constant bonus or price designs with a non-zero bonus. A possible null hypothesis would be H^0_{FTP8}.

H^0_{FTP8} In design B with a proportional bonus, demands are not higher than in the constant bonus or price situations of designs D, E, F, and H.

The respective alternative hypothesis is H^A_{FTP8}:

H^A_{FTP8} In design B with a proportional bonus, demands are higher than in the constant bonus or price situations of designs D, E, F, and H.

Again, a MWU test can be performed, based on the average ranks of the demands for design B (60.63) and the other designs (47.97). This difference between design B and the rest turns out to be significant (MWU-Test, $p =$

0.038, one tailed, Z= -1.77). Based on these test results, H^0_{FTP8} is rejected. The proportional bonus leads to higher demands than other bonus types.

Regarding expected demands, it could be suspected that receivers expect lower demands than receivers are finally willing to make. A possible null hypothesis is H^0_{FTP9}.

H^0_{FTP9} The expected demands are not lower than the real demands.

The respective alternative hypothesis is H^A_{FTP9}:

H^A_{FTP9} The expected demands are lower than the real demands.

To test H^0_{FTP9}, a MWU test is used. The average rank of expected demands s lower (108.23) than the average rank of real demands (147.43). This difference is highly significant (MWU-Test, $p < 0.001$, one tailed, Z = -4.32), leading to the rejection of H^0_{FTP9}. Therewith, the receivers expect lower demands than finally take place. Obviously, they expect to be treated fairer than proposers are willing to be. Design B with a proportional bonus of 10% produced the highest average demands, but also the highest average expected demands. As already mentioned above in this paragraph, a proportional bonus appears to sound more attractive to both receivers and proposers than other bonuses, since both players seem to be willing to allocate more money to the proposer in the presence of a proportional bonus than in other settings, maybe trying to compensate the proposer for his missing bonus payment. But considering the monetary amounts which receivers finally obtain as their 10% bonus, other bonus systems are in fact more interesting to receivers. The difference between demands and expected demands might also account for the difference between the frequency of veto power decisions and expected veto power decisions. Since receivers expect lower demands than proposers are finally willing to make, receivers refrain from veto power more often than proposers expect them to do. The impact of that and all of the other effects on the demands is illustrated in the following paragraph.

c) Graphical Illustration of the Demand Decisions

The validity of several effects has been demonstrated by hypotheses testing. The major aspects are illustrated by Figure 30 below.

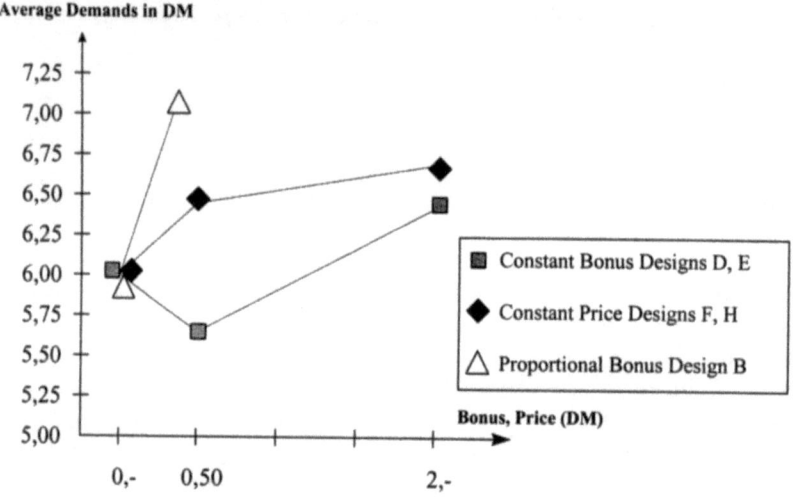

Figure 30: Average Demands for all Bonus Types

The zero bonus design C produces all three symbols at a bonus size of zero, since it can be compared to all proportional bonus, constant bonus and constant price designs. Just like in Figure 28 for the veto power choices above, a bonus value of DM 0,36 is used to represent the 10% bonus of design B. As already discussed during the analysis of hypothesis H^A_{FTP6}, a positive bonus has a certain impact on the demands. Apart from design D, the demands are higher in situations with a bonus. This effect is reinforced by a higher bonus, see H^A_{FTP7}. Additionally, a proportional bonus leads to higher demands than constant bonuses, see H^A_{FTP8}. Neglecting the outcome of design D, a significant difference between the bonus payment and the price payment mechanism does not seem to exist. Altogether, the effect of a bonus on the size of the demands appears to be a little weaker and also less clear than on the veto power decisions. This is illustrated by the decisions as shown in Figure 30 and also by the test results. Further aspects are discussed in chapter G.IV., where a first summary for the FTP game is given.

Obviously, the demands for different bonus situations could be compared to demands in other Dictator, Ultimatum, Impunity, and similar games. It could be expected that the character of the FTP game produces demands in the range of previously observed Ultimatum data, since the FTP game is closer to an Ultimatum than to a Dictator game, because it includes a rejection option. As summarized in chapter B.II.3., Ultimatum games showed a strong tendency towards a two third demand, while Dictator games usually produced average demands in a range between 75 and 90%. In the FTP game, the average demands for the six relevant designs are in a range between 56 and 70%, with an overall average demand of 63%. This is also illustrated by Figure 30 above. Therewith, the outcomes for the range and average of the demands are similar to findings of previous Ultimatum games. To confirm this finding, a statistical test could be performed. The distribution of demands for a previous Ultimatum game could be compared to the distribution of demands in the FTP game. But this does not appear to give further insights towards the behavior in the FTP game, and therefore it is refrained from further testing at this point. Apart from the average demand, the importance of equal splits is also similar to previously reported Ultimatum game results with a frequency of 30%, also being the modal offer. In Dictator games, equal splits usually played a less dominating role. The rejection frequency is also in line with previous findings, averaging 19%. The acceptance decisions are examined in the following paragraph.

3. The Acceptance Decisions

In the third step of the FTP game, more than 80% of the demands of the proposers were accepted by the receivers, as shown by Figure 31. The third step could only be reached with a VP decision in the first step of the game. Due to the low number of rejections in all designs, it is pointless to search for any significant differences between designs.

However, it appears to be possible to explore the characteristics of accepted and rejected demands. The accepted demands were in a range between DM 9,99 and DM 4,50, averaging DM 5,69 ($\sigma = 1,06$). The rejected demands were in a range between DM 10 and DM 5,25, averaging DM 7,42 ($\sigma = 2,02$). A popular test approach in this case is to find out whether rejected demands were higher than accepted ones.

Design	VP Scenarios	No. of Rejections	Rejections in %
A	7	0	0
B	4	1	25
C	17	4	23.5
D	4	1	25
E	2	1	50
F	9	2	22.2
H	5	0	0
Sum	48	9	18.75

Figure 31: Distribution of Rejected Offers

The null hypothesis is H^0_{FTP10}:

H^0_{FTP10} The rejected demands are not higher than the accepted demands.

The respective alternative hypothesis is H^A_{FTP10}:

H^A_{FTP10} The rejected demands are higher than the accepted demands.

To perform a test for H^0_{FTP10}, all rejected demands over all designs are pooled and so are all accepted demands, and a MWU test is performed. On that basis, the difference between accepted and rejected demands is significant (MWU-Test, $p = 0.002$, one tailed, $Z = -3.08$), leading to the rejection of H^0_{FTP10}. Therewith, rejected demands were in fact higher than accepted demands. The efficiency of the different designs was already analyzed in chapter G.I.4. above, using the realized percentages of the maximum payoffs. Due to more VP choices and the resulting higher absolute number of rejections, the no bonus design C was the design with the lowest payoff efficiency. Out of 20 demands, 4 were rejected in design C, while only 5 of 120 demands were rejected in the other designs.

IV. General Results of the FTP Game

Since freedom of choice was the main objective for the implementation of the FTP game, the first discussion of the results should be held in the light of the theory of freedom of choice, which is done in the following paragraph. After that, a more general survey is presented, taking other outcomes, influences, and explanations into account.

1. Interpretation of the Behavior Towards Freedom of Choice

The game Freedom to Punish offers an alternative that can be excluded by the players. This is where freedom of choice can be applied. The veto power can easily be sold for a bonus. Veto power and the resulting punishment option as such do not generate a monetary value. The only meaning of veto power might be a better strategic position, but in this game the strategic position cannot be observed by the proposer and is therefore worthless. The rational decision is to sell the veto power and take advantage of the bonus therewith.

By selling their secret veto power, receivers give up some freedom of choice. The value of this monetary useless alternative of keeping veto power is reflected by the bonus. Depending on the validity of the corresponding hypotheses H^A_{FTP1}, H^A_{FTP2}, and H^A_{FTP3}, conclusions about the relevance of the concept of freedom of choice can be drawn. The most important hypothesis for this study is H^A_{FTP1}. If the relevance of a bonus for the behavior in the FTP game is proven, the relevance of freedom of choice receives some strong support. The other hypotheses also add some support, mainly because they prove the robustness of this effect. The validity of H^A_{FTP1} has been proved by means of statistical testing. There is a clear difference between the zero bonus and the bonus settings with regards to the frequency of veto power choices. A little bonus already led to a strong effect, since a lot of receivers traded their veto power. Over all bonus designs, 76% of the participants sold their veto power. These receivers were confronted with bonus offerings of DM 0.50 and 2.00, but also with an extra average payoff of only DM 0.36 in the proportional bonus setting. In the design without a bonus, only 15% of the participants were willing to exclude their veto power. All others wanted to keep the alternative of a punishment by a rejection of the offer.

There is a price at which players trade unwanted alternatives. This price might be very low or even the smallest possible unit. But nevertheless it has to be positive. This supports the importance of freedom of choice and is therefore contrary to standard (consumer) theory. A more general summary of the experimental results is given in the next paragraph, since freedom of choice should only be seen as one possible interpretation.

2. Overall Outcomes of the FTP Game

In a normal Ultimatum game, punishing the proposer by rejecting the offer leads to a payoff of zero and is therefore economically uninteresting. A rejection of the offer is especially useless when a one-shot game is played, since there are no learning effects. In the FTP game, this is reinforced by the fact that proposers do not even know whether veto power exists or not. In the presence of a bonus, 76% of the receivers refrained from veto power. This effect proved to be significant according to several tests. A bonus lead to more exclusions of veto power than no bonus, and a high bonus lead to more exclusions than a low bonus. The robustness of these results was confirmed by the usage of several different bonus types and sizes.

The high frequency of veto power sales is remarkable, since the bonus is never high enough to fully compensate receivers who have been offered a very low share of the cake. In previous Ultimatum game experiments, receivers frequently rejected shares as high as 40% of the cake or even higher, therewith giving up substantial amounts of money. In the FTP game, receivers seem to be willing to settle for less, since they refrain from veto power to obtain a small monetary compensation. By doing so, they exclude the possibility to punish an unfair proposer. Such a punishment might be caused by considerations of fairness, but still produces an inefficient outcome due to the resulting zero payoffs. It was shown that designs with a bonus lead to a higher level of efficiency than the design without a bonus, since fewer rejections took place. In this context, the difference between decisions before the start of a game, i.e. by programming a strategy, and decisions in a real game with spontaneous play becomes obvious. When given the opportunity to sell veto power, most of the receivers willingly agree since they understand that veto power does not generate a higher payoff, but so does the bonus. In another traditional Ultimatum game situation with spontaneous play, they might decide otherwise and even reject amounts that are higher than such a bonus. Considering these results, it may not be assumed that approaches like the strategy method necessarily lead to the same results as a spontaneous game procedure. The strategy method in general, and also the veto power decision in the FTP game, avoid the exploitation of receivers to the annoyance that might arise in case of greedy demands by proposers. And the high number of veto power choices in the zero bonus design C demonstrates the high willingness of receivers to punish unfair behavior despite any considerations of efficiency.

Since 50% of the proposers expected veto power to be abandoned by receivers over all bonus designs, the high number of equal splits and moderate demands is remarkable. The overall average demand of DM 6,31 shows that the FTP game did not produce higher demands than Ultimatum games, even though the existence of veto power was uncertain. Nevertheless, receivers expected lower demands than finally took place. The bonus also had an effect on demands, since demands were usually higher in the presence of high bonuses. The missing endowment effect analyzed in connection with H^0_{FTP4} is remarkable, but requires further investigation. Altogether, the most striking result is the consequent selling of veto power by receivers in case of a small monetary reward. The results of the FTP game will be compared to the results of the following RAP game in the final summary. Now, the outcomes for the RAP game are demonstrated in the following chapters.

H. Experimental Design for the RAP Game

The design for the RAP game focuses on the impacts of different monetary incentives on the behavior of the players, especially the proposer. Since only two distributions of the cake are available, it is possible to explore how the demand behavior of proposers is influenced by the exact size of the shares that are available to him in different designs. According to a theory of a crowding-out of intrinsic motivation, the fairness of the proposer could be influenced by the payoff combinations that are available to him. More specifically, the proposer should expose less fairness in situations with higher monetary incentives. Another influence on the demands derives from the existence of veto power. These aspects are illustrated by a short introduction to the different designs.

In a first design, the proposer is facing a fair equal division and a division that clearly favors the proposer and is therefore called greedy. Since the proposer is informed about the veto power decision of the receiver, the frequency of fair divisions in situations with and without veto power can be compared. In a second design, a greedy division and a very greedy division are available. The very greedy division allocates nearly the whole cake to the proposer. Again, the difference between demands in veto power and in no veto power situations is interesting, and the impact of the different available allocations leads towards a possible crowding-out. Proposers might choose the allocation that is best for them more frequently the higher their resulting payoff becomes.

A second dimension is added by the bonus. Since the bonus is paid to both players, the proposer is rewarded by a trustful decision of the receiver, namely refraining from veto power. This might lead to reciprocal behavior, i.e. offering the allocation which results in the higher payoff for the receiver. Two bonuses are implemented, a high and a low bonus, which might also result in different behavioral patterns. To complete the experimental design, several subgames are isolated and played. The veto power decision of the receiver is removed, and the remaining subgames are implemented using the same parameters. Therewith, a cardinal Ultimatum and a cardinal Dictator game were played to compare the outcomes with those of the complete games. A comparison of the RAP game results with the results of the FTP game is also important. The different designs are explained in more detail in the following chapter H.I., and the hypotheses are developed in chapter I.II. below.

I. Design Approach for the Experiment

Figure 32 shows the realized experimental sessions for the RAP game, labeled I to IV for the complete games and IU, ID, IIU, and IID for the subgames, where IU represents the Ultimatum subgame of design I and ID the corresponding Dictator subgame. The subgames of designs III and IV were not implemented.

No.	Design	Date	Description	Observations	No. of Persons	Sum of Payoffs
9	I	29.01.98	RAP FGS	8	16	104,-
10		4.05.98		12	24	206,-
11	II	29.01.98	RAP VGS	9	18	163,-
12		4.05.98		11	22	142,-
13	III	4.05.98	RAP FGH	20	40	391,-
14	IV	4.05.98	RAP VGH	20	40	480,-
15	IU	30.11.98	Ultimatum FGS	10	20	140,-
16	ID	30.11.98	Dictator FGS	10	10	130,20
17	IIU	30.11.98	Ultimatum VGS	10	20	140,-
18	IID	4.05.98	Dictator VGS	5	5	87,15
19		30.11.98		4	4	67,20
			TOTAL:	119	219	2.050,55

Figure 32: Design Overview for the RAP Game

All designs were realized using the strategy method. All of the participants had to produce a complete description of their behavior, based on five questions. By means of this method, it becomes possible to let all receiver strategies play against all proposer strategies. Some disadvantages of the strategy method were already discussed. For details about the strategy method, see Selten (1967).

H. Experimental Design for the RAP Game

1. Treatment Variables

The RAP game has the following parameters, which can be varied to observe and isolate certain aspects of decision behavior.

C   The size of the pie is constant.

y   The size of the smaller demand y and therewith the given distributions can be varied. One pair of distributions would be "fair vs. greedy" (y vs. Y). A second pair contains "greedy vs. very greedy" (y* vs. Y*). This parameter should be the main variable for trying to analyze intrinsic motivation.

Δy This parameter is constant. It determines the difference between the two possible distributions within a distribution pair: $Y = y + \Delta y$ and $Y^* = y^* + \Delta y$.

Y   This parameter is varied according to y, since $Y = y + \Delta y$ and $Y^* = y^* + \Delta y$.

δ The bonus can also be varied. Two values should be realized, a high bonus δ^* and a small bonus δ.

The values of the parameters for the RAP sessions will be as follows:

C   = 20 DM
Δy = 6 DM
y   = 10 DM (and therefore Y = 16 DM)
y*   = 13 DM (and therefore Y* = 19 DM)
δ = 5 %
δ^* = 50 %

The difference between the two games FTP and RAP was already discussed. The RAP game offers public information about the veto power choice of the receiver, a bonus for both players in case of no veto power and furthermore is a cardinal game with only two possible demands. But the designs for these games show some differences as well. Apart from the fact that the RAP game is played using the strategy method, some parameters have other values. The size of the cake was DM 10 for the FTP game, but DM 20 for the RAP game. Both values should generate a comparable motivation. For the FTP game, various

I. Design Approach for the Experiment

different bonus types were used. In the RAP game, only a proportional bonus is implemented. The sizes of this bonus with 5% and 50% should result in somewhat different bonus amounts than the low and high bonus of DM 0,50 and DM 2,00 in the FTP game. Finally, the bonus is always paid as bonus, since the price payment design is only used in the design of the FTP game.

Therefore, the following 2 x 2 design structure with four different designs I to IV is established according to Figure 33.

Parameter	δ	δ^*
y	Design I	Design III
y*	Design II	Design IV

Figure 33: Design Structure for the RAP Game

The designs will be referred to using the above numbers and the following abbreviations:

Design I: FGS (fair-greedy, small bonus)
Design II: VGS (greedy-very greedy, small bonus)
Design III: FGH (fair-greedy, high bonus)
Design IV: VGH (greedy-very greedy, high bonus)

These designs are described in more detail in the following paragraphs.

2. Design I with a Small Bonus, a Fair and a Greedy Distribution

In this design, the proposer faces a fair and a greedy distribution of the DM 20 cake. Both players receive a small bonus of 5% in case the receiver abandons his veto power. Figure 34 shows a simplified payoff table, where the acceptance of the receiver is implied. Of course, in case the receiver had chosen VP in the first step, he can reject the offer (leave – L). This leads to a payoff of zero for both players. Refer to Figure 35 for the game in extensive form.

Proposer Receiver	Y		y	
VP,T	4	16 10		10 10
NV	4,2	16,8	10,5	10,5

Figure 34: Payoff Table for Design I

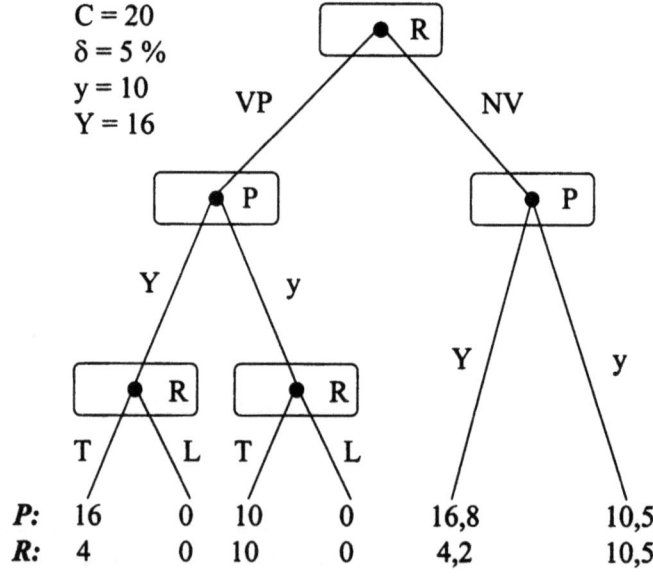

Figure 35: Game-Tree with Parameters of Design I with $\delta = 5\%$

I. Design Approach for the Experiment

3. Design II with a Small Bonus, a Greedy and a Very Greedy Distribution

For design II, the bonus for abandoning veto power is kept at 5%, but the available distributions are changed according to Figure 36, which again does not consider rejections.

Proposer Receiver	Y	y
VP,T	19 1	13 7
NV	19,95 1,05	13,65 7,35

Figure 36: Payoff Table for Design II

The complete game in extensive form is shown by the following Figure 37.

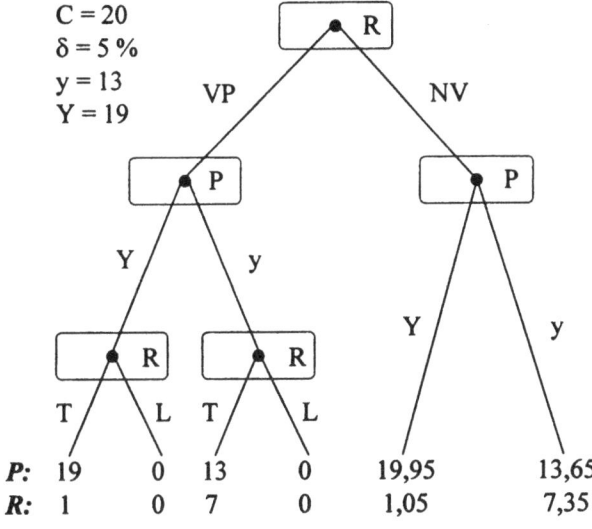

Figure 37: Game-Tree with Parameters of Design II with $\delta = 5\%$

4. Design III with a High Bonus, a Fair and a Greedy Distribution

Designs III and IV both include the usage of a high bonus. 50 % is considered to be high enough to have a significantly different influence than 5%. The impact on the payoffs is illustrated by Figure 38, again not considering rejections.

Proposer Receiver		Y		y	
VP,T	4		16 10		10
NV	6		24 15		15

Figure 38: Payoff Table for Design III

The game in extensive form is shown by Figure 39.

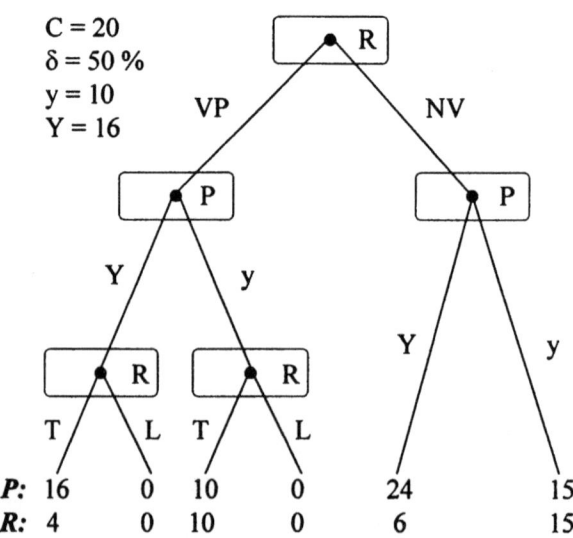

Figure 39: Game-Tree with Parameters of Design III with $\delta = 50\%$

5. Design IV with a High Bonus, a Greedy and a Very Greedy Distribution

In design IV, proposers can realize the highest payoff due to a bonus of 50% and the available very greedy distribution as shown by Figure 40, which again does not consider rejections.

Proposer Receiver		Y	y
VP,T	1	19 7	13
NV	1,5	28,5 10,5	19,5

Figure 40: Payoff Table for Design IV

The complete game in extensive form is shown by the following Figure 41.

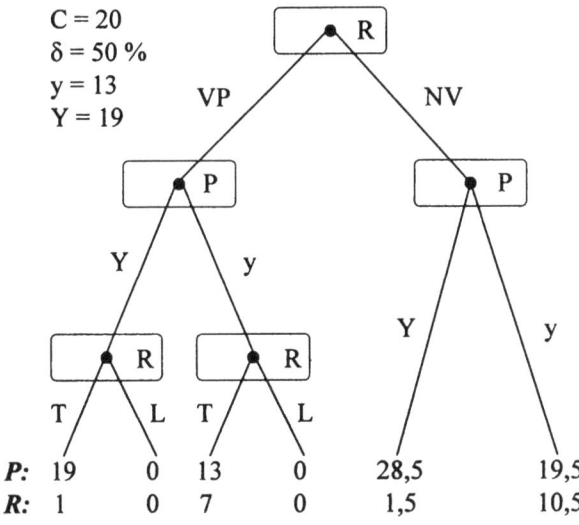

Figure 41: Game-Tree with Parameters of Design IV with $\delta = 50\%$

6. Playing a Subgame

The RAP game can also be played in a reduced form. The first step of the game is eliminated, and the players are confronted with a subgame, which is a rather typical Ultimatum or Dictator game. Of course, the same parameters have to be used and both parts, Ultimatum and Dictator, have to be played to be able to compare the outcomes with those of the complete games. All designs can be used for that purpose, and the new design codes are as follows.

Design ID (or FGSD) for the Dictator-Version of design I

Design IU (or FGSU) for the Ultimatum-Version of design I

The game tree for design ID is displayed in the following Figure 42.

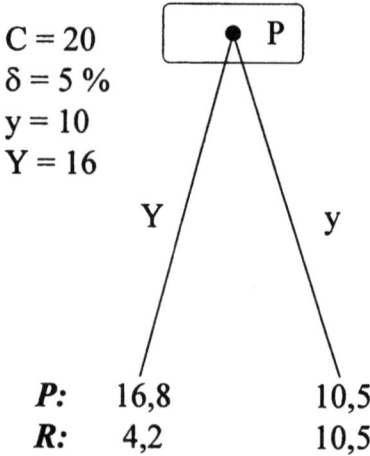

Figure 42: Game-Tree for Design ID

Basically, design ID is a cardinal Dictator game, since the receiver did not abandon his veto power as he could have done in designs I to IV. He is simply assigned the receiver role, and therefore he does not have the possibility to reject the offer. This changes in the subgame design IU, which is a cardinal Ultimatum game as can be seen in Figure 43 below.

II. Alternative Designs

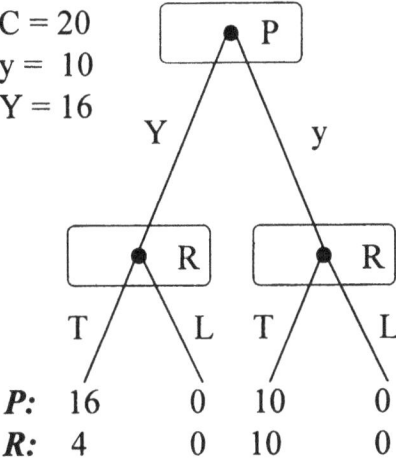

Figure 43: Game-Tree for Design IU

Since veto power exists, there is no bonus. Along with designs ID and IU, the subgame designs for design II, namely IID and IIU were also realized. The respective game trees are similar to those of designs ID and IU, since the only difference is the payoff structure. The subgames of designs III and IV were not implemented. In the next chapter, some design alternatives are discussed. After that, the experimental results of the RAP game are presented.

II. Alternative Designs

For the RAP game, the design options are numerous. The realized designs have been explained, but some additional approaches should be described in this chapter. These should only serve as a reference or inducement for future research. Instead of using the strategy method, this game could also be played as a real three stage game with spontaneous decisions. The results might be different, as players are confronted with offers directly from their matched partner. This might lead to a more emotional response than a programmed strategy. For example, in spontaneous play bad offers might be rejected more often.

Another idea is playing the same game with all possible cake shares. Proposers would not only choose between two given distributions, but could offer any distribution of the cake they want to. The most interesting aspect here would be the difference between offers in the veto power and the no veto power setting, since proposers are informed about the veto power decision of the receivers.

The impacts of the size of the bonus could be analyzed in more detail, either by paying a very high bonus of 100% or by paying a constant bonus like in the FTP game. In this case, receivers could be sure to receive a certain sum and might refrain from veto power more often. But the directions of these effects are of course unclear and would have to be explored. As already described in the design alternatives for the FTP game, the positions of the players could be auctioned, as well as veto power. And the usage of more detailed questionnaires is also possible for the RAP game, since questions of fairness, trust, reciprocity, and intrinsic motivation might play a key role. But before any of the alternative designs should be considered, the results of the original RAP game sessions have to be reviewed, and this is done in the next chapters.

I. Experimental Results for the RAP Game

In this chapter, the experimental results for the RAP game are analyzed. In the first part I., the outcomes of the different decisions are simply listed in an appropriate format to give an overview and to gain a first impression. The following paragraphs II. and III. contain the statistical analysis, and especially some hypotheses testing. After that, the main outcomes are compiled and discussed in paragraph IV.

I. An Overview of the Decisions in the RAP Game

In the RAP game, each player programmed a strategy. It is possible to analyze the whole strategy and its success against other strategies, but it might also be important to examine decisions of single steps over all strategies, for example the veto power decisions. That is done in this chapter, where the decisions of each of the three steps are shown on an aggregated basis as well as split by designs, also including the expectations about the moves of the other player. This is followed by the data of the subgames. After that, the strategy tournament is illustrated and the success of the single strategies is determined, as well as the efficiency of the four different designs. This chapter is intended to provide a first impression of the decision data. The statistical analysis, which follows in chapter III., provides further insights into certain regularities and shows whether these first impressions can be proven by means of statistical testing.

1. The Veto Power Decisions

The strategies include a veto power choice of the receiver. The outcomes of the four designs are somewhat different, as shown in the following Figure 44.

I. Experimental Results for the RAP Game

Design	Veto Power	No Veto Power	Veto Power Expected	No Veto Power Expected
Design I	10	10	11	9
Design II	14	6	14	6
Design III	13	7	13	7
Design IV	10	10	12	8
Sum	47	33	50	30

Figure 44: Realized and Expected VP Decisions for all RAP Designs

In 47 of 80 cases (including designs I to IV), the receiver has chosen veto power. The remaining 33 receivers refrained from their right to punish, therewith choosing NV. The expectations are pretty similar to that. For all designs, at least half of the receivers kept their veto power, with an overall total of 59%, as shown by the following Figure 45. Right after that, Figure 46 demonstrates that veto power was expected just a little more frequently than it finally occurred.

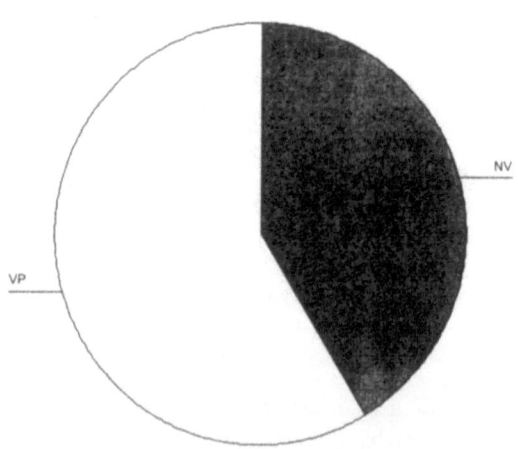

Figure 45: Veto Power Decisions for all RAP Designs

I. An Overview of the Decisions in the RAP Game 129

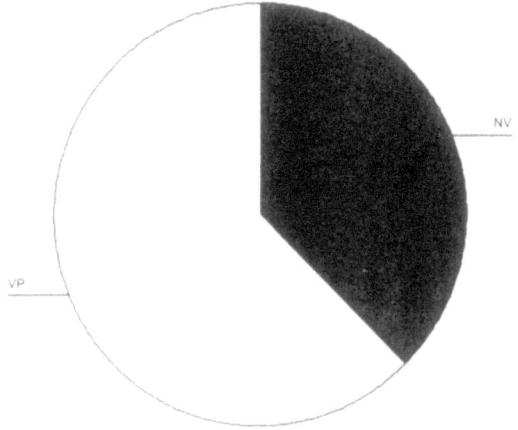

Figure 46: Expected Veto Power Decisions for all RAP Designs

2. The Proposals

The following Figure 47 lists the proposal decisions divided into veto power and no veto power situations. Only two demands were possible, the high demand Y and the low demand y.

Design	Veto Power		No Veto Power		Sum
	Y	y	Y	y	
Design I	10	10	19	1	40
Design II	5	15	20	0	40
Design III	8	12	18	2	40
Design IV	6	14	19	1	40
Sum	29	51	76	4	160

Figure 47: Proposal Decisions for all RAP Designs

In case VP was selected by player R, most of the proposers (64%) demanded only the smaller share y. In the NV scenario, nearly all of the proposers (95%) wanted to have the big share Y. This difference is shown by the following Figure 48 and Figure 49.

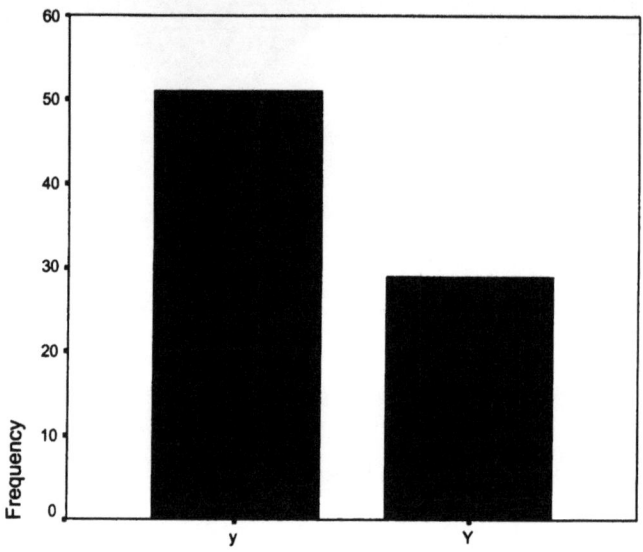

Figure 48: Demands in the Veto Power Situation

The expectations about the demands are included in Figure 50 below. The difference between the veto power and the no veto power setting also exists here, but the total number of Y demands was underestimated by receivers in both situations.

The receivers expected more high demands in the NV than in the VP situation, which is illustrated by Figure 51 and Figure 52 below. Nevertheless, the final demands by proposers in the experiment were greedier for both settings.

I. An Overview of the Decisions in the RAP Game   131

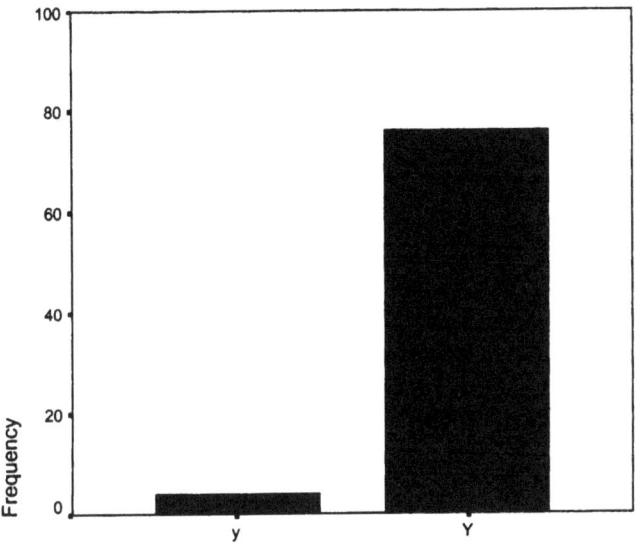

Figure 49: Demands in the Situation Without Veto Power

Design	Veto Power		No Veto Power		Sum
	Y	Y	Y	y	
Design I	3	17	13	5	38
Design II	4	16	13	7	40
Design III	4	16	17	3	40
Design IV	3	17	15	5	40
Sum	14	66	58	20	158

Figure 50: Expected Proposal Decisions for all RAP Designs

132   I. Experimental Results for the RAP Game

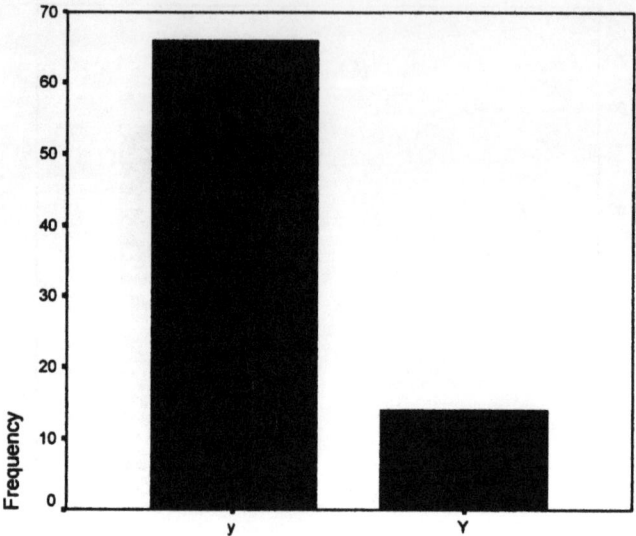

Figure 51: Expected Demands in the Veto Power Situation

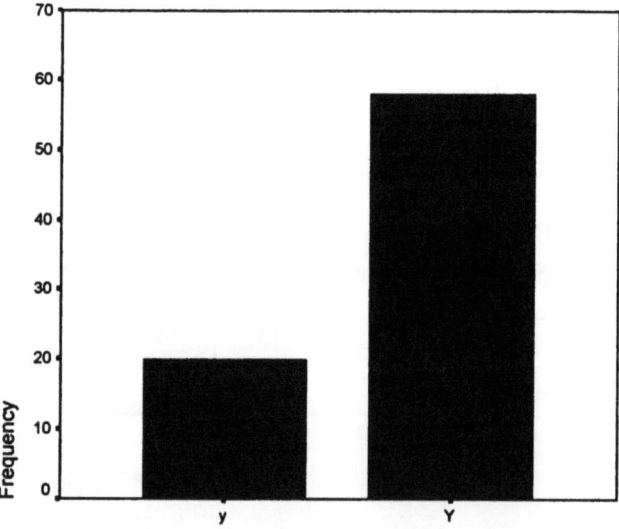

Figure 52: Expected Demands in the Situation Without Veto Power

I. An Overview of the Decisions in the RAP Game   133

Due to these differences between demands and expected demands, some receivers might have been tempted to reject the proposed division. The acceptance decisions are analyzed next.

3. The Acceptance Decisions

Decisions about the acceptance of a proposal only happen in the veto power setting, which was chosen 47 times. As might have been expected, the lower demand y was accepted in nearly all cases, since this is the best offer the receiver can expect (see Figure 53).

Design	Accepted y-demands	Rejected y-demands	Accepted Y-demands	Rejected Y-demands
Design I	10	-	3	7
Design II	13	1	3	11
Design III	13	-	3	10
Design IV	10	-	4	6
Sum	46	1	13	34

Figure 53: The Acceptance Decisions

This changes when Y was chosen by P. In this case, most of the offers were rejected by the receivers (34 of 47 or 72%), as illustrated by the following Figure 54 below.

Similar to these decisions, the proposers also expected the majority of high demands to be rejected, as shown by Figure 55.

The expectations are shown in more detail in Figure 56 below. Not surprisingly, the small demand was expected to be accepted by all but one of the proposers.

134 I. Experimental Results for the RAP Game

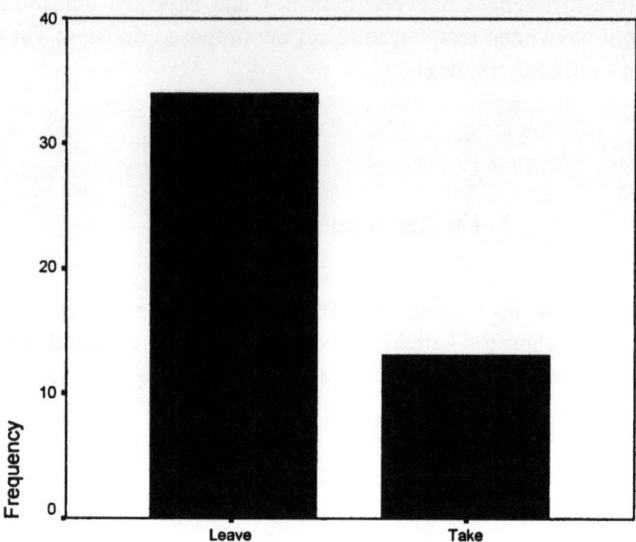

Figure 54: Acceptance Decisions for High Demands

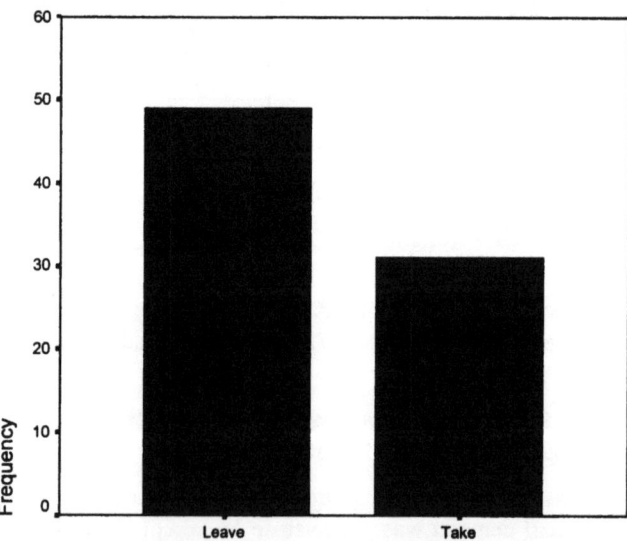

Figure 55: Expected Acceptance Decisions for High Demands

I. An Overview of the Decisions in the RAP Game 135

Design	Expected Accepted y-demands	Expected Rejected y-demands	Expected Accepted Y-demands	Expected Rejected Y-demands
Design I	20	0	11	9
Design II	20	0	7	13
Design III	20	0	7	13
Design IV	19	1	6	14
Sum	79	1	31	49

Figure 56: The Expected Acceptance Decisions

4. The Subgames

The subgames ID, IID, IU, and IIU were played. Of course, there is no veto power decision. The proposals are shown by Figure 57.

Design	Demand		Expected Demand	
	Y	y	Y	y
Design ID	4	6	-	-
Design IID	5	4	-	-
Design IU	4	6	3	7
Design IIU	4	6	4	6
Sum	17	22	7	13

Figure 57: Proposal Decisions for all RAP Subgame Designs

Neither the proposals nor the expectations seem to differ significantly between designs. The acceptance decisions are included in the next Figure 58. Due to the missing veto power, the subgames ID and IID have no third stage and therefore no decision by the receiver at all. Expectations were also not recorded.

I. Experimental Results for the RAP Game

Design	Accepted y-demands	Rejected y-demands	Accepted Y-demands	Rejected Y-demands
Design IU	10	0	4	6
Design IIU	9	1	3	7
Sum	19	1	7	13

Figure 58: The Acceptance Decisions

As already observed for designs I to IV, the low demand was accepted in nearly all of the cases, while the high demand was rejected frequently. The expectations of the proposers about the acceptance behavior of the receivers seem to follow a similar pattern and are included in Figure 59.

Design	Expected Accepted y-demands	Expected Rejected y-demands	Expected Accepted Y-demands	Expected Rejected Y-demands
Design IU	9	1	4	6
Design IIU	8	2	3	7
Sum	17	3	7	13

Figure 59: The Expected Acceptance Decisions

Again, the majority of proposers expected the low demand to be accepted. The opinion about the acceptance of the high demand was mixed. In the next chapter, the possible behavior and player types for the strategy tournament are introduced. After that, the results of the strategy tournament are reported.

5. Behavior Types for Proposers and Receivers

Each proposer is confronted with one design only. Nevertheless it might be interesting to know how he would behave in a different design. Considering the two different distributions and their two possible demands (fair-greedy vs. greedy-very greedy), proposers as well as receivers could match with four

I. An Overview of the Decisions in the RAP Game   137

different behavior types. The bonus is not examined in this context. The following behavior types are possible, see Figure 60. The resulting decision of the proposer appears in a black circle, which is a demand value for the proposer. These types include both the VP and the NV scenario, since they only refer to the way the cake is split. Therewith, strategic considerations are left aside for the most part and only the impact of fairness and other influences on the allocation of the cake matter.

The fair proposer always tries to realize equal shares or shares as equal as possible.	(10) 16 | (13) 19
The rational proposer always demands as much as possible.	10 (16) | 13 (19)
The either 100% fair or 100% rational proposer chooses the equal split only when exactly equal shares are available.	(10) 16 \ 13 (19)
The self-conscious and somewhat fair proposer wants to take advantage of his better position, but also wants to leave a reasonable part of the cake for the receiver.	10 (16) / (13) 19

Figure 60: Behavior Types for the Proposer

I. Experimental Results for the RAP Game

A similar thought experiment can be conducted for the receivers. Again, the behavior in the two different designs fair-greedy and greedy-very greedy is anticipated for a certain type of receiver. The resulting decision about veto power (V) or no veto power (N) in the two different settings appears in a black circle in Figure 61.

The distrustful receiver wants to threat with his position and maybe even reject the high demand.	$\boxed{V_{16}^{10}}$ $\boxed{V_{19}^{13}}$	N_{16}^{10} N_{19}^{13}
The rational receiver wants to realize the bonus. This could also be the trustful receiver, who believes in the fairness of the proposer.	V_{16}^{10} V_{19}^{13}	$\boxed{N_{16}^{10}}$ $\boxed{N_{19}^{13}}$
The nearly rational receiver wants to realize the bonus in the fair-greedy setting, but wants to reject DM 1 offers in the greedy-very greedy setting.	V_{16}^{10} $\boxed{V_{19}^{13}}$	$\boxed{N_{16}^{10}}$ N_{19}^{13}
The somewhat trustful receiver expects no very greedy demands of DM 19 and chooses NV. But in the fair-greedy design, he prefers to have veto power.	$\boxed{V_{16}^{10}}$ V_{19}^{13}	N_{16}^{10} $\boxed{N_{19}^{13}}$

Figure 61: Behavior Types for the Receiver

The relevance of these behavior types can be determined by means of the following strategy tournament. The strategy type that is played most frequently in the fair-greedy scenario and the one that is chosen most frequently in the greedy-very greedy scenario can be combined to one of the behavior types listed above, one for proposers and one for receivers. To be able to determine such types, the frequencies of certain strategies are analyzed in the next chapter.

6. A Strategy Tournament

A strategy tournament can be conducted using the collected decision data. The participants were told whether they were in the role of proposer or receiver. Each player had to answer five questions. For the proposer, the answers to two of the questions were decisions. One question asked about the share he would offer in case of veto power and the next about the share he would offer in case of no veto power. The other three questions inquired about his expectations concerning the decisions of the receiver. For the receiver, three of the answers were decisions. First, whether he would play VP or NV, second, if he would take or leave the small share in case of veto power and third, if he would take or leave the big share in case of veto power. The other two questions regarded his expectations about the decisions of the proposer.

For the strategy tournament, the analysis can be reduced to the observed player types. These types represent all players who programmed the same strategy. Not all types will be examined, since not all of the possible strategies were programmed. A player type will be represented by a short notation of the decisions made by him. For example, the notation "PyYDI" means that the proposer "P" chooses the low demand "y" in case of VP and the high demand "Y" in case of NV and has been playing design I "DI". The player type "RVTLDIII" means that the receiver "R" chooses veto power "V", accepts (takes) the low demand "T", rejects (leaves) the high demand "L", and has been playing design III "DIII". Of course, a player type represents a distinctive strategy as well.

The frequency of the programmed strategies for the proposer is shown by the following Figure 62. Blank cells indicate that the respective strategy was not programmed in the corresponding design. The proposers preferred to play especially two strategies. PyY was played 48 times, and PYY 28 times. Pyy was chosen 3 times, and PYy was played once. PYY is the perfectly rational choice according to game theory. P always demands as much as possible, because R has to accept it anyway to gain money at all. Nevertheless, PYY is not the strategy that was chosen most frequently.

PyY is the strategy that most of the proposers have selected. It anticipates some of the actually observed behavior of R, where the big demand is often rejected in case of veto power. Even though the better offer is made in the veto power scenario, PyY cannot be described as fair. The demand y is only made for strategic reasons, i.e. to avoid a rejection. In the no veto power case, the greedy demand Y is made.

Strategy	Design I	Design II	Design III	Design IV	Sum
Pyy	1		1	1	3
PyY	9	15	11	13	48
PYy			1		1
PYY	10	5	7	6	28
Sum	20	20	20	20	80

Figure 62: Strategies of the Proposer

Pyy was played three times. This is a very fair strategy. No matter whether VP or NV was chosen by R, P always makes the best possible offer. One proposer played PYy, which is a somewhat surprising strategy. But he might have speculated on the bonus, and therefore demanded the big share Y in case veto power was selected, because he wanted a compensation for the lost bonus. This could be interpreted as a punishment for not playing towards the Pareto outcome by wasting the bonus.

The frequency of the programmed strategies for the receiver is shown in the following Figure 63. Strategies implying no veto power are played 33 times, while veto power is chosen 47 times. Frequently chosen strategies are RVTL (33), RNTT (17), RNTL (15) and RVTT (13). Of course, all of the strategies starting with RN imply the same outcome of the game (RNTT, RNTL, RNLT, RNLL). In the case of no veto power, the receiver is not allowed to decide between T and L in a third stage – he has to take the offer. This means that it is irrelevant (especially for the programmed strategy tournament) to differ between these RN strategies. But there are some differences between some of the designs regarding the frequencies of the chosen RN-strategies that might be of interest. Therefore, the RN-strategies are not aggregated.

Strategy	Design I	Design II	Design III	Design IV	Sum
RNTT	6	3	5	3	17
RNTL	4	3	1	7	15
RNLT			1		1
RNLL					-
RVTT	3	3	3	4	13
RVTL	7	10	10	6	33
RVLT					-
RVLL		1			1
Sum	20	20	20	20	80

Figure 63: Strategies of the Receiver

RNTT (and considering the identical outcome, also all of the other RN-type strategies) is the perfectly rational choice according to game theory, since R takes advantage of the bonus. Playing RNTT is a very consequent strategy and was played 17 times. If someone accepts each offer in case of veto power, why should he not play NV and get the bonus? While all of these 17 players R answered the questions about taking or leaving the small or the big share in case of veto power both with Take, some other receivers who played RNTL or RNLT seemed to have a different view of the situation.

15 receivers would not have accepted the higher demand in case of veto power. RNTL is chosen most frequently in the greedy-very greedy design IV. This is especially understandable because of the size of the small share in design IV, which was extremely low at only DM 1,-. The question is why they have chosen an NV strategy and therewith given up their right to reject the small share. It could be suggested that they expected the better offer or simply speculated on the 50 % bonus, even though it is only worth DM 0,50 in design IV. Nevertheless, this behavior is not only triggered by the high bonus for design IV, since this bonus also applies for design III and there does not seem to be a similar tendency in design III. Only the combination of a high bonus and greedy-very greedy shares lead to these decisions.

One receiver programmed RNLT, meaning he would accept the small offer, but not the big offer in the veto power situation. This is inconsistent. But since he did not play veto power, this decision can be ignored. Strategy RNLL was not chosen at all. It implies the same outcome as all RN-strategies, but rejecting both offers in the Ultimatum case, as expressed by LL, would imply a payoff of zero in any case and is therefore difficult to justify.

Two strategies leading to veto power are especially popular, RVTT and RVTL. Both strategies somewhat anticipate the observed behavior of P. In VP, P is very likely to choose the small share y, while he demands the big share Y in case of NV. With respect to this, R can only expect to receive the fair share by choosing VP. This is successful because R's threat to reject the small offer, documented by the VP choice, appears to be strong enough to impress most of the proposers. Despite their veto power choice, the 13 receivers who programmed RVTT accepted both the high and the small offer.

The strategy RVTL also implies the threat to leave the small share by choosing VP in stage 1. But here, the small offer is really rejected, and both players receive a payoff of zero. The receivers are not playing VP to impress P, they do not only threat, they really reject. And RVTL is in fact played far more frequently than RVTT (33 versus 13). Another strategy, RVLT, was never played, while one receiver programmed RVLL, meaning he would not receive a payoff at all. Maybe he was not pleased with his role as receiver, or did not like the experiment at all. But it seems to be more understandable that he was not willing to accept either of the outcomes. Both possible payoff combinations are unfair from his point of view (13:7 or 19:1). And if he anticipated the game-theoretical solution with a payoff of 1,05 for himself, he might have even been frustrated. Additionally, he was confronted with the parameters of design II (greedy-very greedy distribution and small bonus), which are extremely unattractive for the receiver compared to the three other designs. While only 28 proposers have chosen to play the perfectly rational choice PYY, a big number of proposers (52) refrained from playing PYY. This effect is similar, but less strong for the receivers, where the perfectly rational choice NV is played 33 times in contrast to 47 other strategies. Based on the observed strategies, the behavior types for the players can be determined. Figure 64 below shows the frequencies of the strategies PyY and PYY for both designs groups.

Remember that the strategy PyY implies the lower demand in case of veto power. Therefore, the behavior type "fair proposer" seems to be the best description for the observed behavior in case of veto power, even though the difference between PyY and PYY in designs I and III with 20 to 17 is only marginal. As already mentioned in 2 above, the dominating choice in the NV scenario is the big demand, and the corresponding behavior type is the "rational proposer".

I. An Overview of the Decisions in the RAP Game

Strategy	Designs I and III DM 10 and DM 16	Designs II and IV DM 13 and DM 19	Sum
PyY	20	28	48
PYY	17	11	28
Sum	37	39	76

Figure 64: Favorite Strategies for Proposers

Comparing these outcomes, it could be argued that fairness is only pretended to avoid punishment in case of veto power. As shown by Figure 65 below, the frequency of veto power choices does not seem to be heavily influenced by the designs regarding the different available shares.

Strategy implying	Designs I and III DM 10 and DM 16	Designs II and IV DM 13 and DM 19	Sum
Veto Power	23	24	47
No Veto Power	17	16	33
Sum	40	40	80

Figure 65: Favorite Strategies for Receivers

Overall, veto power is chosen more frequently, therewith resulting in the "distrustful receiver" being the dominant behavior type. The strong simplifications for the definition of the behavior types as well as for the selection of the most popular strategies have to be criticized, and the outcomes will be challenged by the statistical analysis in later chapters. The success of the elaborated strategies is explored in the next chapter.

7. Payoffs and Efficiency

The effective payoffs of the players have been determined by randomly matching a strategy of a proposer with one of a receiver. But it is also possible and appears to be more interesting to examine the success of the strategy types in a tournament. In such a tournament, all proposer strategies of a certain design are played against each of the receiver strategies of this design, leading to 20 proposer payoffs.

Design	Strategy	Frequency	Average Payoff	Design Average Payoff
Design I	PyYDI	9	DM 13,40	DM 11,94
	PYYDI	10	DM 10,80	
	PyyDI	1	DM 10,25	
Design II	PyYDII	15	DM 14,44	DM 13,04
	PYYDII	5	DM 8,84	
Design III	PyYDIII	11	DM 14,90	DM 12,95
	PyyDIII	1	DM 11,75	
	PYYDIII	7	DM 10,80	
	PYyDIII	1	DM 7,65	
Design IV	PyYDIV	13	DM 20,75	DM 19,71
	PYYDIV	6	DM 18,05	
	PyyDIV	1	DM 16,25	

Figure 66: Average Payoffs for Proposers

The average payoff of this strategy is used as an indicator for the success of this strategy. As already mentioned above, these figures only serve as first impressions. Any statistical analysis is done in chapter III. The strategies of the proposers are easy to analyze, since there are only four different strategies, and each of these performed similar in all designs. The average payoffs of these strategies are included in Figure 66.

In all four designs, it was better to demand the smaller share y in case of veto power. The higher demand Y was rejected very often (in 72% of the existing veto power situations). In case of no veto power, the proposers nearly always

I. An Overview of the Decisions in the RAP Game   145

offered the smaller share (95%). Proposer payoffs in design IV are somewhat higher due to the overall higher shares for the proposer and the high bonus, which were given by the design specifications. PyY is by far the best proposer strategy in all four designs, followed by PYY. The strategies of the receiver produced the following average payoffs, see Figure 67.

Design	Strategy	Frequency	Average Payoff	Overall Average Payoff
Design I	RVTTDI	3	DM 7,00	DM 5,06
	RVTLDI	7	DM 5,00	
	RNTTDI, RNTLDI	10	DM 4,52	
Design II	RVTTDII	3	DM 5,50	DM 3,77
	RVTLDII	10	DM 5,25	
	RNTTDII, RNTLDII	6	DM 1,05	
	RVLLDII	1	DM 0,00	
Design III	RVTTDIII	3	DM 7,60	DM 6,56
	RNTTDIII, RNTLDIII, RNLTDIII	7	DM 6,90	
	RVTLDIII	10	DM 6,00	
Design IV	RVTTDIV	4	DM 5,20	DM 3,49
	RVTLDIV	6	DM 4,90	
	RNTTDIV, RNTLDIV	10	DM 1,95	

Figure 67: Average Payoffs for Receivers

The best receiver strategy is RVTT in all four designs, followed by RVTL. It is obvious that choosing VP in designs II and IV was by far better than choosing NV, since the average payoffs in case of NV were only DM 1,05 and DM 1,95. In design I, players who have chosen NV had an average payoff of DM 4,52 and DM 6,90 in design III. This difference between design I and II (and also between design III and IV) is generated by the design specifications, since an NV choice, the low offer of DM 1 and the 5 % bonus lead to a payoff of only DM 1,05 in design II, while in design I the low offer of DM 4 and the 5 % bonus amount to DM 4,20.

Surprisingly, the veto power choice RVTL was more successful for the receivers in design II than in design I, even though the sizes of the shares for receivers are smaller in design II than in design I. The overall average payoff for proposers including all designs amounts to DM 14,41, while receivers have to be content with DM 4,72. The efficiency of the different designs can be measured by dividing the average total payoff for both players by the maximum possible payoff. The results are included in Figure 68.

The efficiency of designs I, II, and IV is roughly 80%, while design III produced a far lower efficiency of 65%. This difference is generated by frequent veto power decisions in design III, which lead to a loss of the DM 10,- bonus. Additionally, the total number of rejections is also higher than in other designs.

Design	Average Total Payoff (DM)	Average Possible Payoff (DM)	Efficiency in %
Design I	17,00	21,-	81
Design II	16,81	21,-	80
Design III	19,51	30,-	65
Design IV	23,20	30,-	77
Overall	19,13	25,50	75

Figure 68: Efficiency of the RAP Designs

The following statistical analysis provides further insights into those problems and shows whether these first impressions can be proven by means of statistical testing. In the next chapter, the hypotheses for the RAP game are developed, discussed and tested.

II. Design Background and Hypothesis Approach

The design structure of the RAP game is aimed at analyzing the impact of two variables on the behavior, which are the available share-sizes of the pie and the bonus. In this context, it appears to be worthwhile to determine whether the different available distributions lead to a crowding-out of intrinsic motivation on behalf of proposers. They might be tempted to demand the higher share in certain constellations rather than in others. Especially the very greedy share could crowd out the fairness that might have been applicable in other situations.

But the most interesting aspect is the question whether the behavior of the receivers is different from their behavior in the FTP game, which was analyzed in the previous chapter G. In the FTP game, the veto power decision of the receiver is private information for the receiver, but in the RAP game, this decision is observed by proposers. This difference in the bargaining situation could have a strong impact on behavior. The pattern of the veto power decisions in the RAP game might add another aspect or even proof towards the importance of freedom of choice, but also towards the perception of fairness. Furthermore, the general tendency of the veto power decisions also provides chances to gain further insights into bargaining behavior. Conclusions about the impact of the receiver's possibility to influence the rules of the bargaining process might be drawn as well.

To support any of these thoughts, the approach of hypothesis testing is used, which has already been illustrated in chapter G.II. regarding the testing for the FTP game. The hypotheses are formulated, tested and interpreted in the following chapter III. All null hypotheses referring to the RAP game are labeled H^0_{RAPX}, where X is the number of the hypothesis. Alternative hypotheses are labeled H^A_{RAPX}, respectively. By performing certain statistical tests of these hypotheses, the impact of the influences described above can be evaluated. This is done in the next chapters.

III. Statistical Analysis for the RAP Game

To find some reliable proof for all or at least a few of the characteristics discussed above, several hypotheses are developed and tested in this chapter. Furthermore, some additional effects are illustrated, tested and discussed. The difference between the RAP and the FTP game is tested based on the null hypothesis H^0_{RAP1}. The veto power decisions are examined by means of the hypotheses H^0_{RAP2} and H^0_{RAP3}, the proposals are analyzed by hypotheses H^0_{RAP4}, H^0_{RAP5}, H^0_{RAP6}, H^0_{RAP7}, H^0_{RAP8}, and H^0_{RAP9}. By means of hypothesis H^0_{RAP10}, some conclusions about the acceptance behavior are drawn. Finally, the subgames are considered by hypothesis H^0_{RAP11}.

I. Experimental Results for the RAP Game

1. Differences Between the FTP Game and the RAP Game

The results of the FTP game have been reported in previous chapters. The RAP game offered a bonus for abandoning veto power in all designs, while a bonus of zero was implemented for one of the seven designs of the FTP game. In the FTP game, only 48 of 140 receivers have chosen veto power. In the presence of a bonus for abandoning veto power, only 31 of 120 receivers in the FTP game selected veto power, roughly a quarter of the participants. But in the RAP game, more than half of the receivers (47 of 80) have chosen veto power. Some thoughts should be spent on this difference between the FTP game and the RAP game. In the RAP game, the veto power decision is public knowledge. In the FTP game described in the earlier chapter E., the veto power choice is private information for the receiver. This lead to a remarkably high number of no veto power choices. But in the RAP game, abandoning the veto power results in a very weak position, since this move is observed by the proposer. Therefore, it appears to be plausible that more receivers choose veto power in the RAP game than in the FTP game. To test this effect, a null hypothesis H^0_{RAP1} can be formulated as follows:

H^0_{RAP1} The frequency of veto power choices in the RAP game is not higher than the frequency of VP choices in the FTP game.

The resulting alternative hypothesis is:

H^A_{RAP1} The frequency of veto power choices in the RAP game is higher than in the FTP game.

To perform the testing of H^0_{RAP1}, the Pearson Chi square test of independence is used, see Siegel (1956). This test determines whether two variables are independent. In this case, the independence of the variables "game" and "veto power choice" is tested. This test was described in more detail in chapter G.III.1.b. The difference between the veto power frequencies of the RAP game and the FTP game including all designs is highly significant ($p < 0.001$, $\chi^2 = 12.41$), as well as the difference between those of the RAP game and those of the six bonus designs of the FTP game ($p < 0.001$, $\chi^2 = 21.86$), leading to a rejection of H^0_{RAP1}. Therewith, veto power choices are in fact more likely in the RAP game.

The main difference between the FTP and the RAP game is based on the availability of the information about the veto power choice of the receiver. In the FTP game, the veto power choice is private information for the receiver. The proposer does not know whether the receiver has veto power or not. But in the RAP game, the proposer is informed about the veto power choice before he has to make his offer. Therefore, receivers give up their veto power frequently in the FTP game to take advantage of the bonus, but are more reluctant to refrain from veto power in the RAP game, since they fear to be exploited by the proposer. Private information about the veto power choice leads to more NV choices (76% in the FTP game) than public information (41% in the RAP game). Of course, other differences between the two games might also play a role, for example the fact that the bonus is paid to both players in the RAP game. The different size of the cake could also be relevant in this context, since the higher cake of DM 20 in the RAP game could make veto power more interesting for receivers than the DM 10 cake of the FTP game. But any of these effects is unlikely to be the cause for the impressing difference in veto power choices. The different institutional setting, which leads to deviating information situations, has a strong impact on behavior.

2. The Veto Power Decisions

The veto power choice of the receivers might depend on the available distributions. In the fair-greedy designs I and III, the receivers could be willing to choose NV quite often, and therewith demonstrate trust, since they rely on some intrinsically motivated kind respond on behalf of the proposer. In the greedy-very greedy designs II and IV, this should happen less often because the receivers expect less fairness and therefore want to preserve the possibility to punish the proposers. Therefore, they rather choose VP. The corresponding null hypothesis is H^0_{RAP2}.

H^0_{RAP2} In the fair-greedy designs I and III, NV is not chosen more frequently than in the greedy-very greedy designs II and IV.

The alternative hypothesis to H^0_{RAP2} is H^A_{RAP2}:

H^A_{RAP2} In designs I and III, NV is chosen more frequently than in designs II and IV.

The data shows an effect in this direction, but only for design I and design II. 10 out of 20 receivers have chosen no veto power in design I versus 6 of 20 receivers in design II. Therewith, more receivers in design II tended to keep their veto power to be able to reject very greedy DM 19 demands versus design I. Using the Pearson Chi Square test of independence, this difference between designs I and II is not significant on an acceptable level ($p = 0.197$, $\chi^2 = 1.66$), and therefore the null hypothesis H^0_{RAP2} can not be rejected. Furthermore, designs III and IV show an effect in the opposite direction. Here, the greedy-very greedy design IV motivated more receivers (10) to refrain from veto power than in design III, where only 7 receivers abandoned their right to punish. According to that, the available distributions of the cake do not seem to have a clear effect on the frequency of veto power choices.

Now, the size of the bonus should be considered. While a bonus of 5% does probably not generate a high incentive for abandoning veto power, the high bonus of 50% should sound interesting enough to receivers to consider the NV choice. A possible null hypothesis could be H^0_{RAP3}:

H^0_{RAP3} In designs III and IV with a high bonus, no veto power choices (NV) are not more frequent compared to the low bonus designs I and II.

The alternative hypothesis is:

H^A_{RAP3} In designs III and IV with a high bonus, no veto power choices (NV) are more frequent compared to the low bonus designs I and II.

In this case, a higher bonus would create a crowding in of trust on behalf of receivers. Since the bonus is paid to both players, the receivers behave both strategically and trustful and give the proposers the chance to obtain the high bonus. In return, they expect to be treated fairly and also want to receive the bonus themselves. But a higher bonus might as well crowd out trust (and intrinsic motivation to trust). In case of a low bonus, receivers could be willing to trust the proposers and choose NV, because they expect to be treated fairly. In the presence of a high bonus, receivers could see a higher risk of receiving the smaller share, since the proposers have a high incentive to choose the high share for themselves and obtain a 50% bonus on this share in addition. Therefore, receivers might decide to choose the veto power position.

III. Statistical Analysis for the RAP Game

The experimental evidence is again mixed. While the high bonus in design IV produced less veto power choices than design II with a low bonus (10 versus 14), designs I and III show reverse figures. While only 10 receivers have chosen veto power in design I with a low bonus, the existence of a high bonus for abandoning veto power in design III does not lead to less veto power choices, but to more (13). Using the Pearson Chi Square test of independence, this difference between designs IV and II is not significant on an acceptable level ($p = 0.197$, $\chi^2 = 1.66$), and therefore the null hypothesis H^0_{RAP3} cannot be rejected. Contrary to the assumptions of hypotheses H^A_{RAP2} and H^A_{RAP3}, neither the distributions nor the size of the bonus have a significant straightforward influence on veto power decisions. The pattern of the proposals is analyzed next.

3. The Proposals

In those cases in which receivers have chosen no veto power, proposers might want to take advantage of their strong position by choosing the big share Y instead of the smaller share y. To test this, hypothesis H^0_{RAP4} is formulated.

H^0_{RAP4} In the NV situations, proposers do not chose the big share Y more often than the small share y.

The alternative hypothesis is H^A_{RAP4}:

H^A_{RAP4} In the NV situations, proposers chose the big share Y more often than the small share y.

To test H^0_{RAP4}, a one-sample Chi square test can be performed. This approach tests whether the observed frequencies are different from the expected frequencies. In this case, H^0_{RAP4} implies that the big share Y and the small share y were chosen equally frequent in no veto power situations, which would be 40 times out of the 80 existing cases for the RAP game. Again, such a test can be used in this situation since the number of expected frequencies (40) is clearly above 5. According to the experimental results, the observed frequencies are quite different from those expected frequencies. When NV was chosen, the proposers nearly never offered the better share and played Y in 76 of 80 cases or 95%. This difference between high and low demands (76 to 4) is highly significant ($p < 0.001$, $\chi^2 = 64.80$), leading to the rejection of H^0_{RAP4}. This result shows that fairness in NV situations is hard to find, leaving not much room for a further crowding-out. An examination of crowding-out effects would then have to concentrate on the veto power situations. But before that, the proposals in case of veto power should be analyzed.

In the VP situations, proposers might fear a rejection and therefore rather demand the low share y for themselves. A test could be based on the following null hypothesis H^0_{RAP5}.

H^0_{RAP5} In the VP situations, proposers do not chose the small share y more often than the big share Y.

The alternative hypothesis is H^A_{RAP5}:

H^A_{RAP5} In the VP situations, proposers chose the small share y more often than the big share Y.

To test H^0_{RAP5}, the same procedure based on a one-sample Chi square test is used. Facing veto power, proposers refrained from high demands and chose the low demand more frequently. This difference (51 low demands versus 29 high demands in case of veto power) is also significant ($p = 0.014$, $\chi^2 = 6.05$) and leads to a rejection of H^0_{RAP5}. Due to their fear of a rejection, proposers seem to prefer the smaller demand in case of veto power. To confirm the results for H^0_{RAP4} and H^0_{RAP5}, a comparison between veto power and no veto power proposals appears to be necessary.

Since proposers are informed about the veto power decision, a difference between the demands in veto power situations and in no veto power situations can be suggested. This has already been discussed in connection with H^0_{RAP4} and H^0_{RAP5}. A possible null hypothesis is H^0_{RAP6}:

H^0_{RAP6} In situations with veto power (VP), proposers do not demand the higher share Y less often than in situations without veto power (NV).

III. Statistical Analysis for the RAP Game   153

The respective alternative hypothesis is H^A_{RAP6}:

H^A_{RAP6} In situations with veto power (VP), proposers demand the higher share Y less often than in situations without veto power (NV).

To test H^0_{RAP6}, the Pearson Chi square test of independence is used, see Siegel (1956). This test, which is different from the one-sample Chi square test used in connection with H^0_{RAP4} and H^0_{RAP5}, determines whether two variables are independent. In this case, the independence of the variables "high share demand" and "no veto power choice" is tested. To perform this task, the expected number of cases in each cell is determined, and then compared to the observed number of cases. If the difference between these figures is significant, the null hypothesis can be rejected. The conditions for this test are met, since the number of expected frequencies exceeds 5. As already mentioned above, 76 out of 80 proposers selected Y in case of NV, but in case of VP only 29 of 80 proposers demanded the high share Y. This difference is highly significant ($p < 0.001$, $\chi^2 = 61.20$), leading to a rejection of H^0_{RAP6}. The dominance of low demands in veto power situations and of high demands in no veto power situations appears to be one of the main characteristics of the RAP game data and will be discussed in more detail in chapter I.IV. below.

Altogether, the fair share y was offered 4 times in case of no veto power. 3 of these 4 offers were made in designs with a high bonus. This might be a weak and surely not significant proof for an increase of fairness in case of a higher bonus. Due to the low number of small share demands of only four, a statistical testing procedure is not appropriate. The proposers sometimes honor the NV choice and therewith the bonus with fairness, especially when the bonus is essential. The higher the bonus, the more likely could be a fair offer. But this effect is, at least in this experiment, dominated by other regularities.

The selection of a demand might also depend on the available shares. The greedy-very greedy designs II and IV could produce extreme greediness on behalf of the proposers. In other words, the fairness of the proposers, represented by fair offers, could be crowded out by higher monetary incentives. In the greedy-very greedy design, the cake-shares for the proposers are significantly higher. Therefore the proposers might choose the unfair alternative Y more often in the greedy-very greedy than in the fair-greedy designs. A possible null hypothesis can be formulated as follows:

H^0_{RAP7} In the greedy-very greedy designs (designs II and IV), the high share (Y) is not demanded more frequently than in the fair-greedy designs (designs I and III).

The alternative hypothesis is H^A_{RAP7}:

H^A_{RAP7} In designs II and IV, Y is chosen more frequently than in designs I and III.

On the other hand, the proposers might also feel that the very greedy demand with a distribution of DM 19 for the proposer and DM 1 for the receiver is simply too unfair. Of course, they might just try to avoid a possible conflict and rejection in case of existing veto power. Therefore, they rather choose greedy (DM 13 and DM 7). But in the fair-greedy design, the proposers could feel that a distribution of DM 10 for both is just too fair considering the strength of their position. Therefore, DM 16 to DM 4 might prove to be the favorite choice. This argumentation would suggest that Y is played more frequently in the fair-greedy designs than it is in the greedy-very greedy designs, especially when a veto power decision was made. This would be contrary to hypothesis H^A_{RAP7}.

Regarding H^A_{RAP7}, the evidence is mixed. In situations with veto power, H^A_{RAP7} does not hold, but in no veto power settings the high demand Y is made slightly more often in design II than in design I, and also slightly more often in design IV than in design III. However, the null hypothesis H^0_{RAP7} cannot be rejected. But the demands in no veto power situations allow for some further analysis.

A crowding-out of fairness by higher monetary incentives might be observable in constellations without veto power. In these NV situations, demanding the smaller share can be interpreted as fair behavior, especially since the proposers do not have to fear a rejection. The design of the RAP game allows to check whether this fairness is crowded out by higher monetary incentives, namely by the higher available shares for proposers in the greedy-very greedy designs II and IV. A null hypothesis H^0_{RAP8} reads as follows:

H^0_{RAP8} In situations without veto power, the high share is not demanded more often in the greedy-very greedy designs II and IV than in designs I and III.

III. Statistical Analysis for the RAP Game 155

The alternative statement is:

H^A_{RAP8} In situations without veto power, the high share is demanded more often in the greedy-very greedy designs II and IV than in designs I and III.

The following Figure 69 shows the respective demands.

Design Type	No. of High Demands	Designs with Low Bonus	No. of High Demands	Designs with High Bonus	No. of High Demands
Fair-greedy	37	Design I	19	Design III	18
Greedy-very greedy	39	Design II	20	Design IV	19

Figure 69: High Demands in the Absence of Veto Power

For each pair of designs, the difference between the number of high demands in fair-greedy and in greedy-very greedy situations is only one decision. Therefore, a statistical test is not likely to produce any useful results. Nevertheless, Fisher's exact test was applied for the pooled data of the fair greedy designs versus the greedy-very greedy designs, but showed no significant results ($p = 0.308$, one tailed). Therefore, there is not enough evidence to reject H^0_{RAP8}.

In general, a light crowding out of fairness, represented by low demands, could be suspected to take place, since the high demands are more frequent when proposers can gain more money. The high demand Y leads to a payoff of DM 16 in the fair-greedy settings, but to DM 19 in the greedy-very greedy settings. But with already a high level of high Y demands in designs I and III, there was of course not much room left for further increases. Therefore, there is no significant proof for the hypothesis H^A_{RAP8}.

As already mentioned during the discussion about hypothesis H^A_{RAP7}, the demands in the presence of veto power show some divergent characteristics, see Figure 70. Again, one effect seems to be caused by the different available shares. In fair-greedy settings, the higher share Y could be demanded more often than in the greedy-very greedy settings because the fear of a rejection is higher in case of a very greedy demand.

Design Type	No. of High Demands	Designs with Low Bonus	No. of High Demands	Designs with High Bonus	No. of High Demands
Fair-greedy	18	Design I	10	Design III	8
Greedy-very greedy	11	Design II	5	Design IV	6

Figure 70: High Demands in the Veto Power Situation

A null hypothesis H^0_{RAP9} reads as follows:

H^0_{RAP9} In veto power situations, the high share Y is not demanded more often in the fair-greedy designs I and III than in the greedy-very greedy designs II and IV.

The alternative statement is:

H^A_{RAP9} In veto power situations, the high share Y is demanded more often in the fair-greedy designs I and III than in the greedy-very greedy designs II and IV.

According to Figure 70, the big share Y was chosen in 10 out of 20 or 50% of all cases in design I, while design II produced only 5 high demands (25%). This difference also exists between designs III and IV, but is rather moderate, because Y was chosen in 40% (design III) versus 30% (design IV) of the observed cases. The conditions for the Pearson Chi square test of independence are met. Therefore, this test was applied for the pooled data and proved to be weakly significant ($p = 0.081$, one tailed, $\chi^2 = 2.65$). On a 10% level, hypothesis H^0_{RAP9} can be rejected.

Both design types (low and high bonus) show that proposers are more likely to demand DM 16 (18 times) than DM 19 (11 times). Therewith, a crowding-in of fair behavior takes place, since the DM 3 higher possible payoff of DM 19 in the GVG designs motivated some proposers to demand the low share of DM 13 instead. The question is whether fairness is the reason for this kind of behavior or not. The alternative is simply a fear of rejection. Proposers might anticipate a high rejection rate for DM 19 demands, and therefore prefer to demand less. As will be shown in paragraph I.III.4. below, high demands in the veto power situation were indeed rejected more often than accepted. Possibly, this also led to the surprising result that DM 13 was the demand with the highest frequency (29) in the veto power situation. It only exists in designs II and IV. This demand is closest to two thirds of the cake with 65%. On the other hand, the alternative demand of DM 19 is extremely greedy and therefore very likely to be rejected. Nevertheless, the possible equal split with a DM 10 demand is only chosen 22 times in designs I and III.

An analysis of fair behavior often concentrates on the frequency of equal splits, see chapter C.VII. In their experiments with cardinal Dictator games, Bolton, Katok, and Zwick (1998) show that the frequency of equal splits is roughly the same for cardinal Dictator games and normal Dictator games, i.e. 7 to 15%. Bolton and Zwick (1995) as well as Güth, Huck, and Müller (1998) show a similar effect for cardinal Ultimatum games with 44% and 49% equal or nearly equal splits. In the RAP designs I and III, an equal split of the cake is available. Obviously, the frequency of equal splits in these designs can be compared with the data from the above mentioned previous experiments. It could be expected that the demands in the veto power situation of the RAP game are in the range of previously observed Ultimatum data, because both situations include a rejection option. For the no veto power situation of the RAP game, it is more likely that the demands are similar to those in previous Dictator games. The following Figure 71 lists the percentages of equal splits in designs I and III, distinguishing between situations with and without veto power.

Design	Veto Power	No Veto Power
Design I	50%	5%
Design III	60%	10%

Figure 71: Percentages of Equal Splits

Not surprisingly, the frequencies of equal splits are in line with previous results. To confirm this finding, a statistical test could be performed. The distribution of equal splits in the previous cardinal Ultimatum game of Bolton and Zwick (1995) could be compared to the distribution of equal splits in the veto power situations of the RAP game. But this does not appear to give further insights towards the behavior in the RAP game, and therefore it is refrained from further testing at this point. The same holds for possible comparisons between no veto power situations and cardinal Dictator games. Another observation is that the frequencies of equal splits in no veto power situations are also similar to those reported in studies of the normal (non-cardinal) Dictator game. Despite the cardinal character of the RAP game, it produced comparable percentages of equal splits. This feature of cardinal games was also determined by Bolton, Katok, and Zwick (1998), who compared cardinal Dictator and normal Dictator games. Therewith, this is also not a surprising result. The acceptance decisions are further examined in the following paragraph.

4. The Acceptance Decisions

The analysis of the acceptance decisions is restricted to the 47 cases in which veto power was selected by the receivers. Due to the used strategy method, the acceptance decisions for both demands were recorded. Obviously, high demands were rejected more often than low demands, see Figure 72.

Design	Rejected High Demands in %	Rejected Low Demands in %
Design I	70	0
Design II	79	7
Design III	77	0
Design IV	60	0
Overall	72	2

Figure 72: Percentages of Rejections

The difference between the frequencies of rejected high demands (34 of 47 or 72%) and rejected low demands (1 of 47 or 2%) is plain to see. To confirm the significance of this observation, the following null hypothesis H^0_{RAP10} is formulated:

H^0_{RAP10} In veto power situations, high demands Y are not rejected more frequently than low demands y.

The alternative is:

H^A_{RAP10} In veto power situations, high demands Y are rejected more frequently than low demands y.

To test H^0_{RAP10}, the Pearson Chi square test of independence is used. For the above data, the test is highly significant ($p < 0.001$, $\chi^2 = 49.57$), leading to a rejection of H^0_{RAP10}. Rejections are indeed more likely for the high demand. The rejection rate is especially high in designs II (79%) and III (77%). These two designs also had more veto power choices than designs I and IV. The receivers seem to find it especially important to have a right to punish in designs II and III, and they use it frequently. This is not anticipated by the proposers, as shown by the many rejected high demands. A conflict is somewhat more likely in designs II and III than in I and IV. A possible explanation for the difference between designs I and II is again offered by the extreme DM 19 demand in the greedy-very greedy design II. A share of DM 1 is not attractive enough for R to accept, leading to more rejections than in design I. Surprisingly, this is different concerning designs III and IV. The percentage of receivers who accepted a DM 1 share is higher than the percentage of receivers who accepted DM 4 in design III. However, the differences between any of these designs do not prove to be significant. A last look at the results of the RAP game is done in the following paragraph, which discusses the results for the subgame experiments.

5. The Subgames

The demand structure of the RAP game data reported in chapter I.III.3. above is rather simple, since the low demand dominated the veto power setting and the high demand the no veto power situation. Considering this, it might be useful to compare these results to a control experiment. The subgames IU, ID, IIU, and IID were experimentally examined to find out whether the demands are different without a preceding veto power choice or not. Design IU is basically an Ultimatum game with veto power. Therefore, a certain similarity between design IU and the veto power situations of design I could be suspected. The following Figure 73 shows the distribution of high demands in designs I, II, IU, and IIU.

In fact, the frequency of high demands in the veto power situation of designs I and II (37.5%) is similar to designs IU and IIU with 40%. This could have been expected, and does not provide any further insights. Similar to the Ultimatum subgames, the no veto power situations of design I could be compared with design ID, and design II with IID. Again, it could be suspected that there is no significant difference between these two settings, since both show strong characteristics of a Dictator situation, i.e. no veto power exists.

Design	High Demands in Veto Power Situation	Subgame Design	High Demands in Ultimatum Subgames
Design I	10 of 20	Design IU	4 of 10
Design II	5 of 20	Design IIU	4 of 10
Overall	15 of 40	Overall	8 of 20

Figure 73: High Demands in Designs I, II, IU, and IIU

Surprisingly, design ID only produced 40% high demands, by far less than the no veto power setting of designs I with 95% high demands. Designs II and IID show similar results. A more detailed analysis can be based on Figure 74, which includes designs I, II, ID, and IID.

Design	High Demands in No Veto Power Situation	Subgame Design	High Demands in Dictator Subgames
Design I	19 of 20	Design ID	4 of 10
Design II	20 of 20	Design IID	5 of 9
Overall	39 of 40	Overall	9 of 19

Figure 74: High Demands in Designs I, II, ID, and IID

To test the significance of this effect, the following null hypothesis H^0_{RAP11} will be used:

H^0_{RAP11} In the Dictator subgames ID and IID, the frequency of high demands is not lower than in the no veto power situations of designs I and II.

The alternative hypothesis is

H^A_{RAP11} In the Dictator subgames ID and IID, the frequency of high demands is lower than in the no veto power situations of designs I and II.

To test H^0_{RAP11}, Fisher's exact test is used. For the high demands in design I without veto power and design ID, the difference is highly significant ($p = 0.002$, one tailed). The same holds true for design II versus design IID ($p = 0.005$, one tailed), and of course for the test of the aggregated data with 39 of 40 high demands in designs I and II versus 9 of 19 high demands in the subgames ($p < 0.001$, one tailed), leading to a rejection of H^0_{RAP11}. Even though the payoff structure for both players is identical in design I without veto power and in design ID, the frequencies of high demands differ significantly. The same holds for design II vs. IID. The fair behavior that can be observed in the subgames is nearly completely crowded out by the existence of a preceding veto power decision of the receiver in the RAP game. If a receiver refrained from veto power, the proposer feels invited to demand the maximum possible. In the context of the RAP game, fairness seems to be dominated by the strategic possibilities that are determined by the other player, i.e. the receiver.

IV. General Results of the RAP Game

While the veto power choices have been the main point of interest in the analysis of the previous FTP game, the analysis of the RAP game focuses on the proposals. In addition to that, comparisons of the RAP game with the FTP game and the RAP subgames provide important insights. Since the crowding-out of intrinsic motivation was the main objective for the implementation of the RAP game, the first discussion of the results should be held in the light of the theory of intrinsic motivation, which is done in the following paragraph. After that, a more general survey is presented, taking other outcomes, influences, and explanations into account.

1. Interpretation of the Behavior Towards a Crowding-Out

The results of the RAP game have shown some crowding effects, for example in connection with hypothesis H^A_{RAP9}, which considered demands in the veto power situations. Here, proposers demanded the high share less frequently in case of the greedy-very greedy design. Even though more money was available in this design than in the fair-greedy design, proposers regularly refrained from high demands. The possible crowding-in of fair behavior could alternatively be interpreted as a simple fear of a rejection. Altogether, the crowding effects by higher bonuses or higher available shares appear to have only limited importance. But by implementing other rules for the bargaining process, a different impact of fairness has been observed. In the Dictator subgames, Proposers frequently demonstrated fairness by demanding the low share, as has been illustrated by hypothesis H^A_{RAP11}. In this case, a no veto power decision by the receiver did not take place, but was implied by the experimental setting. In the ordinary RAP game designs, such fairness of proposers is nearly completely crowded out by the preceding veto power decision of the receiver. The proposers seem to have fewer problems in justifying high demands as long as the receivers themselves have chosen to refrain from veto power. If this decision was made by the experimenter, i.e. by conducting a simple Dictator experiment, a certain sense of justice on behalf of proposers does not allow some of them to be greedy. This fairness is strongly crowded out by a preceding no veto power decision of the receiver in the regular three stage RAP game setting.

This institutional difference could also be the explanation for the difference between the RAP and the FTP game results. While proposers in the RAP game feel free to choose the high demand in the absence of veto power, they are in a totally different situation in the FTP game, because they are not informed about the veto power decision. They are especially not informed about a voluntary abstinence from veto power by the receiver like in the case of the RAP game. This is most likely the cause for the effect proven in connection with hypothesis H^A_{RAP1}. In the RAP game, the frequency of veto power choices is high, and this is contrary to the FTP game, where receivers do not have to fear any greediness on behalf of proposers, which could be produced by an observable no veto power decision. The impacts of fairness on behavior do not only depend on the monetary incentives, but also on certain institutional settings.

2. Overall Outcomes of the RAP Game

While the experimental data of the FTP game showed strong effects of the bonus, the RAP game results produced no such outcomes. The results did not change significantly between low and high bonuses. Apart from H^A_{RAP9}, the different allocations did not have a clear impact either. Altogether, the experimental results of the RAP game are quite robust with respect to the used framing methods. Some small effects could be discovered in the pattern of the RAP game data, but they do not prove to be significant, since they are not as clear as comparable effects in the FTP game. The reason might be the dominance of the effect proved by hypotheses H^A_{RAP4}, H^A_{RAP5}, and H^A_{RAP6}, leading to the conclusion that proposers are most likely to demand the high share Y in the absence of veto power and the low share y in case of veto power. This is also enforced by the results of hypothesis H^A_{RAP10}, which has proved that high demands are rejected far more frequent than low demands. Therefore, it does not come as a surprise that even a high bonus of 50% did not attract more receivers to refrain from veto power.

There might be a certain polarization effect produced by the two existing choices of the receiver on the one hand, namely veto power and no veto power, and the two existing choices of the proposer on the other hand, namely low demand and high demand. The proposers might feel that the low demand is the corresponding demand in case of veto power, while the high demand is the respective choice in case of no veto power. The missing significance of other effects simply demands further research and clarification, but this is probably the nature of the experimental approach. The dominance of the polarization effect is a remarkable outcome, and so is its robustness. Again, the institutional setting appears to have a strong impact on behavior.

Possible crowding effects have already been considered in paragraph I.IV.1 above, but the results in connection with hypothesis H^A_{RAP1} are not only important for a crowding-out analysis. Most receivers seem to anticipate the behavior of proposers, and stick to their veto power no matter how high a bonus might be. The fact that the proposer observes the veto power decision is an influence that dominates the behavior of both players. In comparison with the distinct results of the FTP game, this influence represents the main outcome of the RAP game experiments. It ensures that the results of the FTP game are in fact related to freedom of choice rather than other motivations or even just random influences. The following final chapter summarizes the major results of this whole study and gives an outlook on further research possibilities.

J. Summary

The experimental results for Ultimatum and related games were already outlined in chapter B. above, and so were the outcomes of the two new games, FTP and RAP (chapters G. and I.). Therefore, only the major aspects of the new experimental evidence are compiled here, and it is discussed how these results fit into the existing theory. As might have been expected, fairness plays a certain role in the FTP as well as in the RAP game, represented by the frequent appearances of equal splits as well as by the frequent rejections of unfair offers. But fairness did not prove to be the only influence.

In the FTP game, receivers were willing to give up their right to reject offers even for a small bonus amount. This bonus had a strong effect, and a higher bonus generated even more no veto power choices. By considering the remarkably high monetary amounts that are usually rejected in normal Ultimatum game experiments, it appears to be striking that receivers sell their veto power for only DM 0,50 or 5% of the cake. Therewith, they give up the possibility to reject unfair offers and to show their desire for fairness. In the RAP game, the veto power decision was observed by the proposers. Therefore, the majority of the receivers refrained from selling their veto power despite the bonus. Even a high bonus of 50% did not change their behavior. This influence of the institutional setting on behavior is remarkable. A few crowding effects were also discovered in the context of the RAP game. The preceding veto power decision produced the strongest impact on the demands. While fair behavior was frequently observed in the absence of a veto power decision, like in design ID, proposers showed straightforward behavior in case of the preceding veto power decision of the RAP game, namely demanding the high share in case of no veto power and choosing the low share in the presence of veto power. Fairness was crowded out by the new institution of a veto power decision, confirming the importance of the institutional setting for behavior.

Contrary to the few veto power sales in the RAP game, an amazingly high number of receivers refrained from veto power in the FTP game. This significant difference has clearly been attributed to the different information conditions and the existence of a bonus. Most of the receivers are only willing to refrain from veto power if proposers are not informed about this decision. In that case, the receivers do not have to fear to be exploited. Nevertheless, by refraining from veto power, the receivers exclude the possibility to reach equal

J. Summary

zero payoffs by means of a rejection. Since most of the receivers exclude this punishment option for a small bonus, the desire for fairness does not appear to be the main force for a rejection. Instead, the existence of a rejection option serves a strategic purpose, and most of the receivers in the FTP game do not intend to use it and are therefore willing to sell it secretly. The bonus that they have received in return was interpreted in terms of freedom of choice. To formulate general conclusions about a hypothetical price of freedom, some more theoretical work has to be conducted. An alternative experimental implementation should be helpful, maybe based on consumer theory. The existing axiomatic approach by Ahlert and Crüger (1999) could also be implemented and refined if necessary.

It has to be kept in mind that fairness might be stable in some situations, but fragile in others. The influence of institutional settings on fairness may not be ignored. The phenomenon of fairness has to be further investigated, and effects of learning or risk aversion should also be considered in this context. The impacts of social ties and personal communication on fairness might also contribute new insights into bargaining behavior. And finally, the freedom of choice for decision makers should not be underestimated. The research possibilities are numerous, lending plenty of freedom to future scientists.

Bibliography

Abbink, Klaus/*Bolton*, Gary E./*Sadrieh*, Abdolkarim/*Tang*, Fang-Fang (1998): Adaptive Learning versus Punishment in Ultimatum Bargaining, Discussion Paper B-381, University of Bonn.

Abbink, Klaus/*Sadrieh*, Abdolkarim/*Zamir*, Shmuel (1999): The Covered Response Ultimatum Game, Discussion Paper No. B-416, University of Bonn.

Ahlert, Marlies (1993): Freedom of Choice – A Comparison of different Rankings of Opportunity Sets, Social Choice and Welfare 10, 189-207.

Ahlert, Marlies/*Crüger*, Arwed (1999): A Price for Freedom – An Axiomatic and Experimental Approach towards Freedom of Choice, Working Paper, Martin Luther University of Halle-Wittenberg.

Ahlert, Marlies/*Crüger*, Arwed/*Güth*, Werner (2001): How Paulus becomes Saulus – An Experimental Study of Equal Punishment Games, Homo Oeconomicus XVIII, Vol.2, 303-318.

Binmore, K./*Shaked*, A./*Sutton*, J. (1985): Testing Noncooperative Bargaining Theory: A Preliminary Study, American Economic Review 75, 1178-1180.

– (1988): A further Test of Noncooperative Bargaining Theory: Reply, American Economic Review 78, 837-839.

Bohnet, Iris/*Frey*, Bruno S. (1999a): The Sound of Silence in Prisoner's Dilemma and Dictator Games, Journal of Economic Behavior and Organization 38 (1), 43-57.

– (1999b): Social Distance and Other-Regarding Behavior in Dictator Games: Comment, American Economic Review 89, 335-339.

Bolton, Gary E. (1991): A Comparative Model of Bargaining: Theory and Evidence, American Economic Review 81, 1096-1136.

– (1997): The Rationality of Splitting Equally, Journal of Economic Behavior and Organization 32 (3), 365-381.

– (1998), Bargaining and Dilemma Games: From Laboratory Data towards Theoretical Synthesis, Experimental Economics 1 (3), 257-281.

Bolton, Gary E./*Katok*, Elena (1995): An Experimental Test for Gender Differences in Beneficent Behavior, Economics Letters 48, 287-292.

- (1998): An Experimental Test of the Crowding Out Hypothesis: The Nature of Beneficent Behavior, Journal of Economic Behavior and Organization 37 (3), 315-331.

Bolton, Gary E./*Katok*, Elena/*Zwick*, Rami (1998): Dictator Game Giving: Rules of Fairness versus Acts of Kindness, International Journal of Game Theory 27, 269-299.

Bolton, Gary E./*Ockenfels*, Axel (1998): Strategy and Equity: An ERC-Analysis of the Güth-van Damme Game, Journal of Mathematical Psychology 42, 215-226.

- (1999): ERC: A Theory of Equity, Reciprocity and Competition, American Economic Review (forthcoming).

Bolton, Gary E./*Zwick*, Rami (1995): Anonymity versus Punishment in Ultimatum Bargaining, Games and Economic Behavior 10, 95-121.

Bornstein, Gary/*Yaniv*, Ilan (1998): Individual and Group Behvior in the Ultimatum Game: Are groups more "rational" players? Experimental Economics 1, 101-108.

Bossert, Walter/*Pattanaik*, Prasanta K./*Xu*, Yongsheng (1994): Ranking Opportunity Sets: An Axiomatic Approach, Journal of Economic Theory 63, 326-345.

Brosig, Jeannette/*Ockenfels*, Axel/*Weimann*, Joachim (1999): Why Communication Enhances Cooperation, Working Paper, University of Magdeburg.

Burlando, Roberto/*Hey*, John D. (1997): Do Anglo-Saxons Free Ride More?, Journal of Public Economics 64, 41-60.

Camerer, Colin F. (1995): Individual Decision Making, John H. Kagel/Alvin E. Roth (eds.), Handbook of Experimental Economics, Princeton, 587-703.

- (1997): Progress in Behavioral Game Theory, Journal of Economic Perspectives 11, 167-188.

Camerer, Colin F./*Thaler*, Richard H. (1995): Anomalies: Ultimatums, Dictators and Manners, Journal of Economic Perspectives 9, 209-219.

Cameron, Lisa (1995): Raising the Stakes in the Ultimatum Game: Experimental Evidence From Indonesia, Industrial Relations Section Working Paper No. 345, Princeton University.

Crüger, Arwed (1996): Effizienz und Kooperation bei Produktionsentscheidungen – eine computergestützte experimentelle Untersuchung, Marburg.

Crüger, Arwed/*Königstein*, Manfred (1999): A Principal-Agent Game with active Principals, Working Paper, Humboldt University of Berlin.

Davis, Douglas D./*Holt*, Charles A. (1993): Experimental Economics, Princeton.

Deci, Edward L. (1975): Intrinsic Motivation, New York.

Deci, Edward L./*Ryan*, Richard M. (1980): The Empirical Exploration of Intrinsic Motivational Processes, Advances in Experimental Social Psychology 13, 39-80.

Duffy, John/*Feltovich*, Nick (1999): Does Observation of Others Affect Learning in Strategic Environments? An Experimental Study, International Journal of Game Theory 28, 131-152.

Eckel, Catherine C./*Grossman*, Phillip J. (1996a): Altruism in Anonymous Dictator Games, Games and Economic Behavior 16, 181-191.

– (1996b): The Relative Price of Fairness: Gender Differences in Punishment Games, Journal of Economic Behavior and Organization 30 (2), 143-158.

Erev, Ido/*Roth*, Alvin E. (1998): Predicting How People Play Games: Reinforcement learning in Experimental Games with Unique, Mixed Strategy Equilibria, American Economic Review 88, 848-881.

Everitt, B.S. (1977): The Analysis of Contingency Tables, London.

Falk, Armin/*Gächter*, Simon/*Kovacs*, Judit (1998): Intrinsic Motivation and Extrinsic Incentives in a Repeated Game with Incomplete Contracts, Working Paper, University of Zurich.

Fehr, Ernst/*Gächter*, Simon (1998): Reciprocity and Economics: The Economic Implications of Homo Reciprocans, European Economic Review 42, 845-859.

Fehr, Ernst/*Kirchler*, Erich/*Weichbold*, Andreas/*Gächter*, Simon (1998): When Social Norms Overpower Competition: Gift Exchange in Experimental Markets, Journal of Labor Economics 16, 1998, 321-354.

Fehr, Ernst/*Kirchsteiger*, Georg/*Riedl*, Arno (1998): Gift exchange and reciprocity in competitive experimental markets, European Economic Review 42, 1-34.

Fehr, Ernst/*Schmidt*, Klaus M. (1999): A Theory of Fairness, Competition, and Cooperation, Quarterly Journal of Economics 114 (3), 817-868.

Forsythe, Robert/*Horowitz*, Joel/*Savin* N.E./*Sefton*, Mart (1994): Fairness in Simple Bargaining Experiments, Games and Economic Behavior 6, 347-369.

Frey, Bruno S. (1994): How Intrinsic Motivation is Crowded Out and In, Rationality and Society 6, 334-352.

– (1997a): From the Price to the Crowding Effect, Swiss Journal of Economics and Statistics 133, 325-350.

– (1997b): A Constitution for Knaves crowds out civic Virtues, The Economic Journal 107, 1043-1053.

– (1997c): Not just for the money – An Economic Theory of Personal Motivation, Cheltenham.

Frey, Bruno S./*Bohnet*, Iris (1997): Identification in Democratic Society, Journal of Socio-Economics 26, 25-38.

Frey, Bruno S./*Eichenberger*, Reiner (1994): Economic Incentives transform psychological Anomalies, Journal of Economic Behavior and Organization 23, 215-234.

Frey, Bruno S./*Oberholzer-Gee*, Felix (1997): The Cost of Price Incentives: An Empirical Analysis of Motivation Crowding-Out, American Economic Review 87, 746-755.

Frey, Bruno S./*Oberholzer-Gee*, Felix/*Eichenberger*, Reiner (1996): The Old Lady visits your Backyard: A Tale of Morals and Markets, Journal of Political Economy 104, 1297-1313.

Frey, Bruno S./*Osterloh*, Margit (1997): Sanktionen oder Seelenmassage ? Motivationale Grundlagen der Unternehmensführung, Die Betriebswirtschaft 57, 307-321.

Fudenberg, Drew/*Tirole*, Jean (1991): Game Theory, Cambridge.

Gaertner, Wulf (1990): On Professor Sen's capability approach, Nussbaum, Martha/Sen, Amartya K. (eds.), The quality of life, Oxford.

Gale, John/*Binmore*, Kenneth G./*Samuelson*, Larry (1995): Learning to be Imperfect: The Ultimatum Game, Games and Economic Behavior 8, 56-90.

Geanakoplos, John/*Pearce*, David/*Stacchetti*, Ennio (1989): Psychological Games and Sequential Rationality, Games and Economic Behavior 1, 60-79.

Gravel, Nicolas (1994): Can a Ranking of Opportunity Sets Attach an Intrinsic Importance to Freedom of Choice ?, American Economic Review 84, 454-458.

Greene, William H. (1993): Econometric Analysis, Englewood Cliffs.

Güth, Werner (1992): Spieltheorie und ökonomische (Bei)Spiele, Berlin.

– (1995): On Ultimatum Bargaining Experiments – A Personal Review, Journal of Economic Behavior and Organization 27, 329-344.

Güth, Werner/*Huck*, Steffen (1997): From Ultimatum Bargaining to Dictatorship – An Experimental Study of four Games varying in Veto Power, Metroeconomica, 262-279.

Güth, Werner/*Huck*, Steffen/*Müller*, Wieland (1998): The Relevance of Equal Splits – On a Behavioral Discontinuity in Ultimatum Games, Discussion Paper No. 7, SFB 373, Humboldt University of Berlin

Güth, Werner/*Kliemt*, Hartmut (1997): Intrinsische Motivation: Ausnahme oder Regel? Die Betriebswirtschaft 57, 585-586.

Güth, Werner/*Ockenfels*, Peter/*Tietz*, Reinhard (1992): Distributive Justice versus Bargaining Power, S.E.G. Lea/P. Webley/B. M. Young (eds.), New Directions in Economic Psychology: Theory, Experiment and Application, 153-175.

Güth, Werner/*Ockenfels*, Peter/*Wendel*, Markus (1993): Efficiency by Trust in Fairness? Multiperiod Ultimatum Bargaining Experiments with an Increasing Cake, International Journal of Game Theory 22, 51-73.

Güth, Werner/*Schmittberger*, Rolf/*Schwarze*, Bernd (1982): An Experimental Analysis of Ultimatum Bargaining, Journal of Economic Behavior and Organization 3, 367-388.

Güth, Werner/*Tietz*, Reinhard (1986): Auctioning Ultimatum Bargaining Positions, Roland W. Scholz (ed.), Current Issues in West German Decision Research, Frankfurt, 173-185.

– (1990): Ultimatum Bargaining Behavior – A survey and comparison of experimental results, Journal of Economic Psychology 11, 417-449.

Güth, Werner/*van Damme*, Eric (1998): Information, Strategic Behavior and Fairness in Ultimatum Bargaining: An Experimental Study, Journal of Mathematical Psychology 42, 227-247.

Güth, Werner/*Yaari*, Menahem E. (1992): Explaining Reciprocal Behavior in Simple Strategic Games: An Evolutionary Approach, Ulrich Witt (ed.), Explaining Process and Change – Approaches to Evolutionary Economics, Ann Arbor, 23-34.

Harrison, Glenn W./*Hirshleifer*, Jack (1989): An Experimental Evaluation of Weakest Link/Best Shot Models of Public Goods, Journal of Political Economy, 97 (1), 201-225.

Harrison, Glenn W./*McCabe*, K.A. (1996): Expectations and Fairness in a Simple Bargaining Experiment, International Journal of Game Theory 25, 303-327.

Harrison, Glenn W./*McKee*, Michael (1985): Experimental Evaluation of the Coase Theorem; Journal of Law and Economics 28; 653-670.

Harsanyi, John C./*Selten*, Reinhard (1988): A General Theory of Equilibrium Selection in Games, Cambridge.

Hoffman, Elizabeth/*McCabe*, Kevin/*Shachat*, Keith/*Smith*, Vernon (1994): Preferences, Property Rights and Anonymity in Bargaining Games, Games and Economic Behavior 7, 346-380.

Hoffman, Elizabeth/*McCabe*, Kevin/*Smith*, Vernon (1996a): On Expectations and the Monetary Stakes in Ultimatum Games, International Journal of Game Theory 25, 289-302.

– (1996b): Social Distance and Other-Regarding Behavior in Dictator Games, American Economic Review 86, 653-660.

- (1999): Social Distance and Other-Regarding Behavior in Dictator Games: Reply, American Economic Review 89, 340-341.

Hoffman, Elizabeth/*Spitzer*, Matthew L. (1982): The Coase Theorem: Some Experimental Test; Journal of Law and Economics 25, 73-98.

- (1985): Entitlements, Rights and Fairness: An Experimental Examination of subjects' concepts of distributive justice, Journal of Legal Studies 14, 259-297.

Kagel, John H./*Kim*, Chung/*Moser*, Donald (1996): Ultimatum Games with Asymmetric Information and Asymmetric Payoffs, Games and Economic Behavior 13, 100-110.

Kagel, John H./*Roth*, Alvin E. (1995): Handbook of Experimental Economics, Princeton.

Kahnemann, Daniel/*Knetsch*, Jack L./*Thaler*, Richard (1986a): Fairness as a Constraint on Profit-Seeking: Entitlements in the Market, American Economic Review 76, 728-741.

- (1986b): Fairness and the Assumptions of Economics, Journal of Business 59, 285-300.

Königstein, Manfred/*Tietz*, Reinhard (1994): Profit Sharing in an Asymmetric Bargaining Game, Working Paper, Frankfurt.

Kreps, David M. (1997): Intrinsic Motivation and Extrinsic Incentives, American Economic Review 87, 359-364.

- (1979): A representation theorem for "Preferences for Flexibility", Econometrica 47, 565-577.

Kreps, David M./*Wilson*, Robert (1982): Sequential Equilibria, Econometrica 50, 863-894.

Lindbeck, Assar (1997): Incentives and Social Norms in Household Behavior, American Economic Review 87, 370-377.

- (1988): Individual Freedom and Welfare State Policy, European Economic Review 32, 295-318.

Luce, R. Duncan/*Raiffa*, Howard (1957): Games and Decisions, New York.

Nash, John F. (1950): The Bargaining Problem, Econometrica 18, 155-162.

- (1951): Non-Cooperative Games, Annals of Mathematics 54, 286-295.

Nehring, Klaus/*Puppe*, Clemens (1996): Continuous Extensions of an Order on a Set to the Power Set, Journal of Economic Theory 68, 456-479.

Ochs, Jack/*Roth*, Alvin E. (1989): An Experimental Study of Sequential Bargaining, American Economic Review 79, 355-384.

Ockenfels, Axel/*Weimann*, Joachim (1999): Types and Patterns: An Experimental East-West-German Comparison of Cooperation and Solidarity, Journal of Public Economics 71 (2), 275-288.

Pattanaik, Prasanta K./*Xu*, Yongsheng (1990): On Ranking Opportunity Sets in Terms of Freedom of Choice, Recherches Économiques de Louvain 56, 383-390.

Prasnikar, Vesna/*Roth*, Alvin E. (1992): Considerations of Fairness and Strategy: Experimental Data From Sequential Games, Quarterly Journal of Economics 107, 865-888.

Puppe, Clemens (1995): Freedom of Choice and Rational Decisions, Social Choice and Welfare 12, 137-153.

– (1996): An Axiomatic Approach to "Preference for Freedom of Choice", Journal of Economic Theory 68, 174-199.

Rabin, Matthew (1993): Incorporating Fairness into Game Theory and Economics, American Economic Review 83, 1281-1302.

Rasmussen, Eric (1989): An Introduction to Game Theory, Cambridge.

Rawls, John (1971): A Theory of Justice, Oxford.

Roth, Alvin E. (1995): Bargaining Experiments, John H. *Kagel*/Alvin E. *Roth* (eds.), Handbook of Experimental Economics, Princeton, 253-348.

Roth, Alvin E./*Prasnikar*, Vesna/*Okuno-Fujiwara*, Masahiro/*Zamir*, Shmuel (1991): Bargaining and Market Behavior in Jerusalem, Ljubljana, Pittsburgh, and Tokyo, American Economic Review 81, 1068-1095.

Rubinstein, Ariel (1982): Perfect Equilibrium in a Bargaining Model, Econometrica 50, 97-109.

Schelling, Thomas C. (1960): The Strategy of Conflict, Cambridge.

Schotter, Andrew/*Weiss*, Avi/*Zapater*, Inigo (1996): Fairness and Survival in Ultimatum and Dictatorship Games, Journal of Economic Behavior and Organization 31 (1), 37-56.

Selten, Reinhard (1965): Spieltheoretische Behandlung eines Oligopolmodells mit Nachfrageträgheit, Zeitschrift für die gesamte Staatswissenschaft 121, 301-324 (part I), 667-689 (part II).

– (1967): Die Strategiemethode zur Erforschung des eingeschränkt rationalen Verhaltens im Rahmen eines Oligopolexperiments, Heinz Sauermann (ed.), Beiträge zur Experimentellen Wirtschaftsforschung 1, 136-168.

– (1975): Reexamination of the Perfectness Concept for Equilibrium Points in Extensive Games, International Journal of Game Theory 4, 25-55.

Selten, Reinhard/*Ockenfels*, Axel (1998): An Experimental Solidarity Game, Journal of Economic Behavior and Organization 34 (4), 517-39.

Sen, Amartya (1988): Freedom of Choice, European Economic Review 32, 269-294.

– (1997): Maximization and the Act of Choice, Econometrica 65, 745-779.

Siegel, Sidney (1956): Nonparametric Statistics for the Behavioral Sciences, New York and Tokio.

Simon, Herbert A. (1957): Models of Man, New York.

– (1981): Entscheidungsverhalten in Organisationen, Landsberg am Lech.

Slonim, Robert/*Roth*, Alvin E. (1998): Learning in High Stakes Ultimatum Games: An Experiment in the Slovak Republic, Econometrica 66, 569-596.

Staw, Barry M. (1989): Intrinsic and Extrinsic Motivation, H. Leavitt/L. Pondy/D. Boje (eds.), Readings in Managerial Psychology, Chicago, 36-71.

Straub, Paul G./*Murnighan*, J. Keith (1995): An Experimental Investigation of Ultimatum Games: Common Knowledge, Fairness, Expectations and Lowest Acceptable Offers, Journal of Economic Behavior and Organization 27 (3), 345-364.

Suleiman, Ramzi (1996), Expectations and fairness in a modified Ultimatum game, Journal of Economic Psychology 17, 531-554.

Tietz, Reinhard (1984): The Prominence Standard, Part I, Frankfurter Arbeiten zur experimentellen Wirtschaftsforschung Nr. A 18, Frankfurt.

– (1992): The shift of supply and demand functions due to the spiral of planning, Frankfurter Arbeiten zur experimentellen Wirtschaftsforschung No. A 32, Frankfurt.

– (1999): 25 Jahre BÖMA oder: Ist der Ausstattungseffekt gegen Entscheidungsvorbereitung immun?, Discussion paper, Frankfurt.

Tietz, Reinhard/*Weber*, Hans-Jürgen (1980): Experimentelle Wirtschaftsforschung, Willi Albers et. al. (eds.), Handwörterbuch der Wirtschaftswissenschaften Bd. 2, Stuttgart, 518-524.

van Damme, Eric (1987): Stability and Perfection of Nash Equilibria, Berlin.

de Waal, Frans (1996): Good Natured: The Origins of Right and Wrong in Humans and Other Animals, Cambridge.

Winter, Eyal/*Zamir*, Shmuel (1997): An Experiment with Ultimatum Bargaining in a Changing Environment, Working Paper No. 159, Center for Rationality and Interactive Decision Theory, Jerusalem.

Zajac, Edward E. (1995): Political Economy of Fairness, Cambridge.

Subject Index

Acceptance 66
Annoyance 50
Auction market game 23

Bargaining games 19
Behavior 38
Best Shot game 23
Bonus 67

Cardinal Impunity game 21
Cardinal Ultimatum games 22

Competitiveness 48
Crowding-out 41

Dictator game 20

Efficiency 95
Equal Punishment game 16
Equilibrium strategy 69, 73
Experimental economics 15

Fairness 42

Fairness norm 44
Freedom of choice 58
Freedom to Punish 68

Game tree 80

Impunity games 21
Intrinsic motivation 41

Level of annoyance 52

Payoff 95
Prisoner's Dilemma 24
Proposer 66

Receiver 66
Right and Choice to Punish 71

Social distance 49

Ultimatum game 20

Veto power 67

Printed by Libri Plureos GmbH
in Hamburg, Germany